Practical Calculus

Xing Zhou

Math for Gifted Students

http://www.mathallstar.org

Copyright © 2015 by Xing Zhou. All rights reserved.

No part of this book may be reproduced, distributed or transmitted in any form or by any means, including photocopying, scanning, or other electronic or mechanical methods, without written permission of the author.

To promote education and knowledge sharing, reuse of some contents of this book may be permitted, courtesy of the author, provided that: (1) the use is reasonable; (2) the source is properly quoted; (3) the user bears all responsibility, damage and consequence of such use. The author hereby explicitly disclaims any responsibility and liability; (4) the author is notified in advance; and (5) the author encourages, but does not enforce, the user to adopt similar policies towards any derived work based on such use.

Please visit the website http://www.mathallstar.org for more information or email contact@mathallstar.org for suggestions, comments, questions and all copyright related issues.

use your mobile device to scan this QR code for more resources including books, practice problems, online courses, and blog.

This book was produced using the LaTeX system.

Contents

1 **Introduction**   1

2 **Limit**   3
   2.1 Limit Defined 3
   2.2 Limit Computation 6
   2.3 The Sandwich Theorem 8
   2.4 Bounded Monotonic Function 11
   2.5 Left, Right Limit and Continuity 13
   2.6 Examples and Applications 16
      2.6.1 Finding Asymptotes 16
      2.6.2 Sum Infinite Series 18
      2.6.3 Compute Area Under A Curve 19
      2.6.4 Continuously Compounded Interest 20
   2.7 Practice 22

3 **Derivative**   25
   3.1 Derivative Defined 25
   3.2 Differentiability vs Continuity 27
   3.3 Compute Derivative (I) 28
   3.4 Derivative's Properties 30
   3.5 Derivative of e^x 32
   3.6 Notation Demystification 33
   3.7 Implicit Derivative 34
   3.8 Differentiate Parametric Function 35
   3.9 Compute Derivative (II) 36
   3.10 Inverse Function Rule 37
   3.11 Derivatives of Trigonometric Functions 39
   3.12 The Chain Rule 40
   3.13 The Quotient Rule 43
   3.14 Table of Common Derivatives 44
   3.15 Additional Techniques 45
      3.15.1 Differentiate x^x 45
      3.15.2 Product Rule For Higher Order 46
   3.16 Concavity 47

CONTENTS

    3.17 Partial Derivative 49
    3.18 Examples and Applications. 51
          3.18.1 Determine Minimum and Maximum 51
          3.18.2 Determine Inflection Points 55
          3.18.3 The L'Hôpital Rule 55
          3.18.4 Root Finding Algorithm 57
          3.18.5 Regression and Machine Learning 60
    3.19 Practice 62

4 Integral   **67**
    4.1 Rectangular Approximation Model 67
    4.2 Riemann Sum and Integral 68
    4.3 Fundamental Theorem of Calculus 70
    4.4 The Substitution Method. 74
    4.5 Trigonometric Substitution 76
    4.6 Integration By Parts 78
    4.7 Additional Techniques 80
          4.7.1 Watch Out Absolute Value 81
          4.7.2 The Symmetry Method 82
          4.7.3 The Symmetry Method (II) 83
          4.7.4 Special Pattern 84
          4.7.5 Partial Fraction Decomposition 85
          4.7.6 Completing the Square 87
          4.7.7 The Construction Method 88
          4.7.8 Recursion 90
    4.8 Improper Integral 92
    4.9 Differential Equation 95
          4.9.1 Separable Differential Equation 95
          4.9.2 Integrating Factor 96
          4.9.3 Homogeneous Equation 97
    4.10 Examples and Applications. 99
          4.10.1 Compute Arc Length 99
          4.10.2 Compute Area Using Polar Coordinates . . . 101
          4.10.3 Compute Volume 102
          4.10.4 Compute Surface Area 104
          4.10.5 Determine Center of Mass 106
          4.10.6 Derivative and Integral in Physics 110
          4.10.7 Differential Equation in Physics 111
    4.11 Practice 113

5 Infinite Series   **119**
    5.1 Convergence 119

		5.1.1	Divergence Test 121

        5.1.1 Divergence Test 121
        5.1.2 Comparison Test 122
        5.1.3 Absolute Convergence Test 123
        5.1.4 Ratio Test 124
        5.1.5 Root Test 125
        5.1.6 Limit Test 126
        5.1.7 Integration Test 126
        5.1.8 Alternating Series Test 128
  5.2 Taylor Expansion 129
  5.3 Deriving Taylor Expansion 132
  5.4 Approximation Error 133
  5.5 Examples and Applications 134
        5.5.1 Limit Calculation 134
        5.5.2 Series Differentiation and Integral 135
        5.5.3 Infinitely Nested Radical 136
        5.5.4 Estimation 137
  5.6 Practice 139

Appendices   143

A Solutions   145

  A.1 *Chapter 1* 146
  A.2 *Chapter 2* 147
  A.3 *Chapter 3* 155
  A.4 *Chapter 4* 166
  A.5 *Chapter 5* 181

CONTENTS

Preface

Welcome to Math All Star© series!

Math All Star originates from a series of lectures given to a group of gifted middle school students with a love for mathematics and an interest in participating in competitions such as MathCounts, AMC, and AIME. These lectures aim to strengthen their problem-solving abilities and to introduce effective techniques that are not typically taught in the classroom.

As the popularity of Math All Star grew, the author began to upload lecture materials to create online courses, thereby providing students with the opportunity to progress at their own paces.

Since then, course materials have constantly been reviewed and updated to reflect student feedback and the observations made during lectures. Recent competition problems are also continuously analyzed and referenced to ensure the relevance of the contents. These course materials are the foundations of this Math All Star series.

Because competition math is a diversified subject that covers both a wide breadth and depth of topics, it is quite challenging to effectively cover all the material in one book that is appropriate for every interested student. Consequently, the author has decided to write a series of books, with each one focusing on a particular topic. Students are encouraged to pick and choose where to begin, depending on their individual skill levels and needs.

CONTENTS

In addition to these books, the Math All Star website provides extra practice problems and serves as a highly recommended supplemental learning resource.

If there are any questions, comments, or concerns, please visit the website or email **contact@mathallstar.org**.

Happy learning!

To visit the Math All Star website, scan this QR code or go directly to http://www.mathallstar.org

Chapter 1

Introduction

This book is not intended to replace an 1000-page textbook which gives detailed lectures on every aspect of calculus. Instead, it is intended to explain core concepts in an intuitive way and to focus on practical perspectives of learning calculus.

First, this book emphasizes on computational skills. Most readers will use calculus as a computational tool in various scientific areas such as mathematics, physics, engineering, computer science, quantitative finance, and so on. Therefore, being able to correctly and quickly calculate derivatives, integrals etc is extremely useful and important to these readers. Many calculus theorems are also covered in this book. However, overly abstract and academical details will be kept to a minimal. Such contents are very important to those students who are interested in pursuing pure mathematics. However, focusing too much on these topics may be a distraction to most other students.

Another feature of this book is its focus on relations between calculus and various high school computational techniques such as trigonometric identities, polynomial transformation, and so on. Proficiency in using these techniques plays a vital role in solving many calculus problems. While these methods and formulas will be

Chapter 1: Introduction

included in this book when they are needed, it is strongly recommended that readers should review these topics. Discussion of these techniques can be found in some Math All Star series books such as *Power Calculation*, *Trigonometry*, *Competition Algebra*, and so on.

Meanwhile, this book also includes many examples and applications in various subjects which use calculus. These include continuous compounded interest in finance, root finding algorithm in computer science, optimization in machine learning, and of course various applications in physics. This will help prepare students for applying their calculus skills in their future study and career.

Finally, some selective historical competition problems are also included. These exercises provide students at different levels with an opportunity to test their problems solving skills that suit their individualized learning objective. Full solutions with detailed explanations are included.

More practice problems can be found on our website:

`http://www.mathallstar.org`

Chapter 2

Limit

2.1 Limit Defined

Let's start the journey by examining the behavior of function $y = \frac{1}{x}$ when x becomes infinitely large.

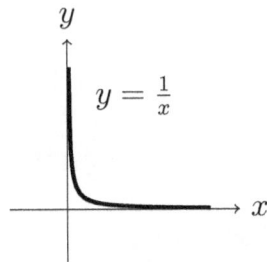

It is intuitive to claim that the value of this function approaches 0 when x becomes infinitely large. This assertion is correct and can be written as
$$\lim_{x \to \infty} \frac{1}{x} = 0$$
where the symbol ∞ means infinity.

Chapter 2: Limit

The notation $\lim_{x \to c} f(x) = L$ means the value of $f(x)$ will become L when x approaches c. It is worth noting that the function $f(x)$ may be undefined at $x = c$. In fact, if $f(x)$ is well defined at the point c, then it becomes trivial to determine that $\lim_{x \to c} f(x) = f(c)$. As a result, the focus of limit calculation is on those cases when the function is undefined at $x = c$.

Because $f(x)$ may be undefined at the point $x = c$, therefore the existence of limit depends on whether or not $f(x)$ will approach a certain value when x approaches c. This is defined as follows:

Definition 2.1.1 **The ε-δ Definition of Limit**

A function $f(x)$ has a limit L when x approaches c if for any arbitrary positive number ε, there exists a number δ such that

$$|x - c| < \delta \implies |f(x) - L| < \varepsilon \qquad (2.1)$$

Because the choice of ε is arbitrary, therefore the value of ε can be as small as $10^{-999999999}$ or even smaller. The relation (2.1) means that regardless of how tiny ε is, the difference between $f(x)$ and L can *always* be smaller if x is sufficiently close to c.

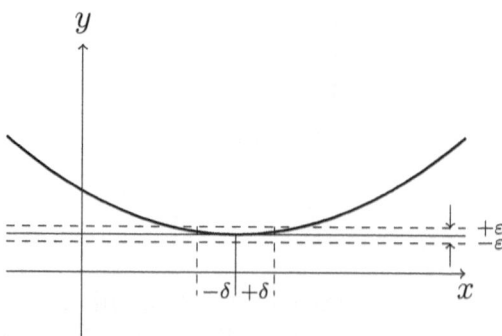

By this definition, whether or not $f(x)$ is defined at $x = c$ is not important as long as the value of $f(x)$ is always within $(L-\varepsilon,\ L+\varepsilon)$

when $x \in (x - \delta, x + \delta)$.

If c is infinite as in the case of $\lim\limits_{x \to \infty} \dfrac{1}{x}$, then the above ε-δ definition is equivalent to

> **Definition 2.1.2 Limit Definition ($x \to \infty$)**
>
> A function $f(x)$ has a limit of L when x becomes infinitely large if for any arbitrary positive number $\varepsilon > 0$, there exists a number X such that
>
> $$x > X \implies |f(x) - L| < \varepsilon$$

In plain English, this means that the value of $f(x)$ can always be as close to L as desired if x is sufficiently large.

Applying *Definition 2.1.1* or *2.1.2* is the first method to prove that a limit exists and also to find the value of the limit. Let's use this method to formally show $\frac{1}{x}$ has a limit of 0 when x becomes infinitely large.

Example 2.1.1

Prove
$$\lim_{x \to \infty} \frac{1}{x} = 0$$

Proof

For any positive number $\varepsilon > 0$, let $X = \frac{1}{\varepsilon}$, then for any $x > X$,

$$0 < \frac{1}{x} < \frac{1}{X} = \varepsilon \quad \text{or} \quad \left|\frac{1}{x} - 0\right| < \varepsilon$$

Then, applying *Definition 2.1.2* leads to the desired claim.

QED

Chapter 2: Limit

2.2 Limit Computation

In addition to applying the basic definition, transforming the given expression to a combination of basic forms is also widely used in limit computation. These transformations may employ polynomial identities, trigonometric identities, and various other calculation techniques. Let's consider a couple of examples.

Example 2.2.1

Evaluate the value of

$$\lim_{n \to \infty} \sum_{k=1}^{n} \frac{1}{k(k+1)} = \frac{1}{1 \times 2} + \frac{1}{2 \times 3} + \frac{1}{3 \times 4} + \cdots$$

Solution

Let

$$S_n = \sum_{k=1}^{n} \frac{1}{k(k+1)} = \frac{1}{1 \times 2} + \frac{1}{2 \times 3} + \frac{1}{3 \times 4} + \cdots + \frac{1}{n \times (n+1)}$$

Applying the telescoping sequence technique as discussed in the book *Power Calculation* gives

$$\begin{aligned} S_n &= \frac{1}{1 \times 2} + \frac{1}{2 \times 3} + \frac{1}{3 \times 4} + \cdots \\ &= \left(\frac{1}{1} - \frac{1}{2}\right) + \left(\frac{1}{2} - \frac{1}{3}\right) + \cdots + \left(\frac{1}{n} - \frac{1}{n+1}\right) \\ &= 1 - \frac{1}{n+1} \end{aligned}$$

As n approaches infinity, $\frac{1}{n+1}$ will approach 0. Hence, we find

$$\lim_{n \to \infty} \sum_{n=1}^{\infty} \frac{1}{n(n+1)} = \lim_{n \to \infty} \sum_{n=1}^{\infty} \left(1 - \frac{1}{n+1}\right) = \boxed{1}$$

Done.

Example 2.2.1 shows how to calculate the sum of an infinite series. As it shows, two steps are required. The first step is to compute the sum of its first n terms, S_n. It is usually an expression of n. Then, the second step is to take n to infinity. The limit is the desired answer.

Example 2.2.2

Evaluate the value of

$$\lim_{n\to\infty} \frac{1}{n} \sum_{k=1}^{n} \left(\frac{k}{n}\right)^2$$

Solution

Let $S_n = \dfrac{1}{n} \sum_{k=1}^{n} \left(\dfrac{k}{n}\right)^2$, then

$$S_n = \frac{1}{n^3} \sum_{k=1}^{n} k^2 = \frac{1}{n^3} \cdot \frac{n(n+1)(2n+1)}{6} = \frac{1}{6} \frac{(n+1)(2n+1)}{n^2}$$

We note that the numerator has two terms each of which is a one-degree polynomial of n, and the denominator is a two-degree polynomial of n. Therefore, we divide each of these one-degree polynomials in numerator by n and have

$$S_n = \frac{1}{6}\left(1 + \frac{1}{n}\right)\left(2 + \frac{1}{n}\right)$$

It follows that

$$\lim_{n\to\infty} S_n = \frac{1}{6} \times 1 \times 2 = \boxed{\frac{1}{3}}$$

Done.

In fact, *Example 2.2.2* is the limit of a Riemann sum which lays the foundation of the integral. Integrals will be studied in *Chapter 4*.

Chapter 2: Limit

2.3 The Sandwich Theorem

A more advanced method to compute limits is to use the squeezing technique which is also referred as the Sandwich theorem. This method can be explained using the diagram below.

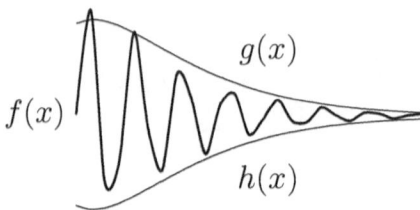

Intuitively, if there exist two functions $g(x)$ and $h(x)$ which have the same limit and wrap around the target function $f(x)$, then it can be asserted that $f(x)$ will converge to the same limit as $g(x)$ and $h(x)$ do.

Theorem 2.3.1   The Sandwich Theorem

If it holds that $g(x) \geq f(x) \geq h(x)$ for $|x - c| < \delta$ where δ is a constant, and
$$\lim_{x \to c} g(x) = \lim_{x \to c} h(x) = L$$
then
$$\lim_{x \to c} f(x) = L$$

It is worth emphasizing that the inequality $g(x) \geq f(x) \geq h(x)$ does not need to hold over the entire domain. It is only required that this inequality holds for all x in the neighborhood of target c.

Example 2.3.1

Show that
$$\lim_{x \to 0} \frac{x}{\sin x} = 1$$

Proof

A common trigonometric inequality states that[1]

$$\sin x \leq x \leq \tan x, \qquad \left(0 \leq x < \frac{\pi}{2}\right)$$

Dividing this relation by $\sin x$ leads to

$$1 \leq \frac{x}{\sin x} \leq \frac{1}{\cos x}$$

When x is negative and $-\frac{\pi}{2} < x < 0$, we have $\sin x \geq x \geq \tan x$. Dividing this relation by $\sin x$ gives (note that in this case $\sin x < 0$, therefore the direction of inequality should be changed.)

$$1 \leq \frac{x}{\sin x} \leq \frac{1}{\cos x}$$

Hence, by setting $\delta = \frac{\pi}{2}$, we have

$$|x - 0| < \delta = \frac{\pi}{2} \implies 1 \leq \frac{x}{\sin x} \leq \frac{1}{\cos x}$$

The left side of the above inequality is a constant independent of x. The right side approaches 1 when $x \to 0$. Therefore, we conclude

$$\boxed{\lim_{x \to 0} \frac{x}{\sin x} = 1} \qquad (2.2)$$

QED

One interesting aspect of *(2.2)* is both its numerator and denominator approach 0 when x approaches 0. Hence, *(2.2)* is to find the limit of a function in the form of $\frac{0}{0}$. It turns out that the limit of such a function may or may not exist and, if the limit exists, it can be any value depending on the relative speeds of the two parts when approaching 0. In the case of $\sin x$ vs x, their speeds are the same, therefore the result equals 1.

---

[1]Readers can refer to the book *Trigonometry* in the *Math All Star* series.

Chapter 2: Limit

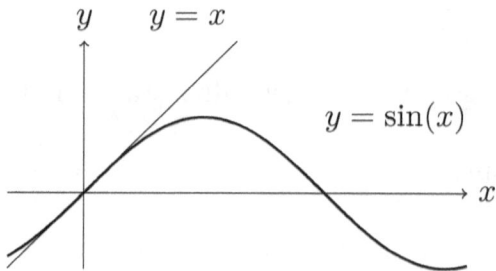

A general and usually easier method to compute the limit of a function in the form of $\frac{0}{0}$ is to apply the L'Hôpital rule. This method will be discussed in the next chapter when derivatives are introduced.

Equation 2.2 is an important relation which is used widely. The example below is an application of this relation. This problem also illustrates the use of expression transformation.

Example 2.3.2

Compute the value of $\lim\limits_{n \to \infty} n^2 \left(1 - \cos \dfrac{\pi}{n}\right)$.

Solution

The given expression can be transformed as

$$\lim_{n \to \infty} n^2 \left(1 - \cos \frac{\pi}{n}\right)$$

$$= \lim_{n \to \infty} n^2 \frac{\left(1 - \cos \frac{\pi}{n}\right)\left(1 + \cos \frac{\pi}{n}\right)}{\left(1 + \cos \frac{\pi}{n}\right)}$$

$$= \lim_{n \to \infty} n^2 \frac{\left(\sin \frac{\pi}{n}\right)^2}{\left(1 + \cos \frac{\pi}{n}\right)}$$

$$= \lim_{n \to \infty} \frac{\pi^2}{\left(1 + \cos \frac{\pi}{n}\right)} \left(\frac{\sin \frac{\pi}{n}}{\frac{\pi}{n}}\right)^2$$

Now, the second term above will approach 1 as n approaches infinitely large by *(2.2)* on *page 9*. The first term will become $\frac{\pi}{2}$ because $\frac{\pi}{n}$ will approach 0. Hence, the final result is $\boxed{\dfrac{\pi^2}{2}}$

Done.

The solution given in this example shows some skillful manipulations. This is not uncommon in calculus computations. Therefore, it is useful to be proficient in various elementary techniques as mentioned in *Chapter 1*. Meanwhile, it is also imperative to realize that there may be multiple ways to solve the same problem. Hence, identifying the most appropriate technique is an important skill to learn. For instance, *Example 2.3.2* can also be solved using the Taylor's expansion. This technique will be discussed in *Chapter 5*.

2.4 Bounded Monotonic Function

If function $f(x)$ is monotonic, then it is possible to determine whether or not the limit of this function exits by examining the function's boundaries.

> **Theorem 2.4.1 Bounded Monotonic Function**
>
> If function $f(x)$ is monotonically increasing with an upper bound, then the limit of $f(x)$ must exist as x approaches infinity. Similarly, the limit of a monotonically decreasing function with a lower bound must exist too.

Similar to the case of Sandwich theorem, $f(x)$ does not need to be to monotonic over its entire domain. It is sufficient that $f(x)$ becomes monotonic when x becomes sufficiently large.

Chapter 2: Limit

Let's take $f(x) = \frac{1}{x}$ as an example. While it is not monotonic in its entire domain, it is monotonically decreasing when $x > 0$ with a lower bound of 0. Therefore, we can assert that $\frac{1}{x}$ will be convergent to a non-negative value as x approaches infinitely large.

Unlike Sandwich theorem, this method can only assert the existence of a limit but cannot determine its value.

Example 2.4.1

Does the following limit exist?

$$\lim_{n \to \infty} \sum_{k=1}^{n} \frac{1}{n^2} = \frac{1}{1^2} + \frac{1}{2^2} + \frac{1}{3^2} + \frac{1}{4^2} + \cdots$$

Solution

This answer is yes. This is because for every positive n, we have

$$\frac{1}{n^2} < \frac{1}{(n-1)n}$$

This means that

$$\frac{1}{2^2} < \frac{1}{1 \times 2}, \quad \frac{1}{3^2} < \frac{1}{2 \times 3}, \quad \frac{1}{4^2} < \frac{1}{3 \times 4} \quad \cdots$$

Therefore

$$\begin{aligned}
& \frac{1}{1^2} + \frac{1}{2^2} + \frac{1}{3^2} + \frac{1}{4^2} + \cdots \\
& < \frac{1}{1^2} + \underbrace{\frac{1}{1 \times 2} + \frac{1}{2 \times 3} + \frac{1}{3 \times 4} + \cdots}_{\text{Example 2.2.1 on page 6}} \\
& = 1 + 1 = 2
\end{aligned}$$

Hence, we find this sum is monotonically increasing as n increases with an upper bound of 2. Therefore, this limit exists and is less than 2. With the help of other techniques, it can be shown that the value of this limit equals $\frac{\pi^2}{6}$.

Chapter 2: Limit

Done.

Similarly, it can be shown that the function $f(n) = \left(1 + \frac{1}{n}\right)^n$ is monotonically increasing and is less than 3. Hence, we can assert that the value below

$$e = \lim_{n \to \infty} \left(1 + \frac{1}{n}\right)^n = \lim_{n \to 0} (1+n)^{\frac{1}{n}} \qquad (2.3)$$

exits. The proof is left as an exercise. This number, $e \approx 2.718$, also referred as Euler's number, is one of the most important constants in calculus and many other scientific subjects.

2.5 Left, Right Limit and Continuity

In this section, let's revisit the function $\frac{1}{x}$ to investigate the case when x approaches 0. As shown in the diagram below, there are two different results depending on the direction from which x approaches 0.

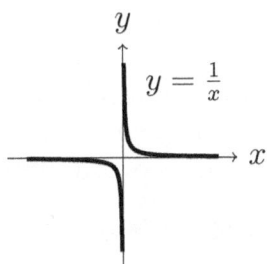

If x approaches 0 by increasing its value (e.g. from the left side of the origin), $f(x)$ will approach $-\infty$. If x approaches 0 by decreasing its value (e.g. from the right side), then $f(x)$ will approach $+\infty$. These two are called the *left limit* and the *right limit*, respectively. They are written as $\lim_{x \to 0^-} \frac{1}{x}$ and $\lim_{x \to 0^+} \frac{1}{x}$.

Chapter 2: Limit

Another example in which the left and right limits do not equal is the floor function $f(x) = \lfloor x \rfloor$. This function returns the largest integer not exceeding the given real number x. Its functional graph, shown below, consists a series of steps where a solid ending point means the function will take this value at the jump and a hollow circle means this value is not achievable. For example when $x = 1$, the function returns 1, not 0.

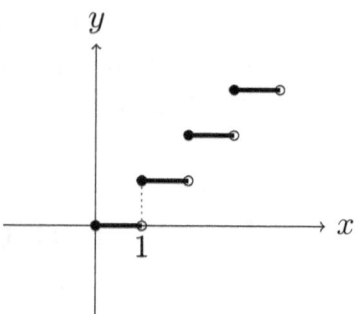

Now, let's consider $\lim_{x \to 1} \lfloor x \rfloor$. From the graph above, it is clear that when x approaches 1 from the left, the result will be 0 and when x approaches 1 from the right, the result will be 1, i.e.:

$$\lim_{x \to 1^-} \lfloor x \rfloor = 0 \quad \text{and} \quad \lim_{x \to 1^+} \lfloor x \rfloor = 1$$

If the left and right limits are not equal, then the "regular" limit does not exist at this point. It also means that this function has a jump at this point, or in another word, is discontinuous here.

Consequently, in order for $f(x)$ to be continuous at point $x = c$, it must hold that $\lim_{x \to c^-} f(x) = \lim_{x \to c^+} f(x)$. However, having equal left and right limits is a necessary but not sufficient condition for a function to be continuous at the said point. A counter example is given below:

$$f(x) = \begin{cases} 0 & , x \neq 0 \\ 1 & , x = 0 \end{cases} \implies \lim_{x \to 0^-} f(x) = \lim_{x \to 0^+} f(x) = 0$$

Despite of having equal left and right limits, this function clearly has an abrupt jump at $x = 0$.

> **Definition 2.5.1 Continuity At A Point**
>
> Function $f(x)$ is continuous at point $x = c$ if and only if
> $$\lim_{x \to c^-} f(x) = \lim_{x \to c^+} f(x) = f(c) \qquad (2.4)$$

Because $\lim_{x \to c^-} f(x) = \lim_{x \to c^+} f(x)$ implies existence of $\lim_{x \to c} f(x)$, therefore *(2.4)* can be simplified as

$$\lim_{x \to c} f(x) = f(c) \qquad (2.5)$$

For function $f(x)$ to be continuous over an interval, it is necessary for $f(x)$ to be continuous at every interior point within this interval by satisfying *(2.4)* or *(2.5)*. However, at the ending points of a closed interval, only one side continuity is required. Let a and b be the left and right ending points, respectively, then $f(x)$ will be continuous at these two points if

$$\lim_{x \to a^+} f(x) = f(a) \quad \text{and} \quad \lim_{x \to b^-} f(x) = f(b)$$

Because an interval consists of a collection of interior points and, possibly, up to two ending points, and a function's domain contains a collection of intervals, therefore

> **Definition 2.5.2 Continuous Function**
>
> A function $f(x)$ is continuous over an interval if it is continuous at every interior point and, possibly, ending point. A function is called a continuous function if it is continuous over its entire domain.

When learning mathematics, it is imperative to pay attention to every detail because math is precise subject. Let's consider a tricky question.

Chapter 2: Limit

Example 2.5.1

Is $f(x) = \frac{1}{x}$ a continuous function?

Solution

The answer is yes. Despite $f(x)$ is discontinuous at the point $x = 0$, this point is not in this function's domain therefore does not need to be considered. Then, because $f(x)$ is continuous everywhere else, we conclude it is a continuous function.

Done.

2.6 Examples and Applications

2.6.1 Finding Asymptotes

Limit is a useful tool to find asymptotes. A function $f(x)$ has a horizontal asymptote $y = C$ if at least one of the following relations holds:
$$\lim_{x \to -\infty} f(x) = C \quad \text{or} \quad \lim_{x \to +\infty} f(x) = C$$

Similarly, $f(x)$ has a vertical asymptote $x = C$ if at least one of the following relations holds:
$$\lim_{x \to C^-} f(x) = \pm\infty \quad \text{or} \quad \lim_{x \to C^+} f(x) = \pm\infty$$

Using limits to find vertical asymptotes is generally more difficult than to determine horizontal ones. One useful tip to is to examine those points where $f(x)$ are undefined. They are often related to vertical asymptotes. Another approach is to first determine this function's inverse and then find the horizontal asymptotes of $f^{-1}(x)$. The horizontal asymptotes of $f^{-1}(x)$ are the vertical asymptotes of $f(x)$.

Let's consider two examples.

Example 2.6.1

Determine the asymptotes of function $f(x) = \frac{1}{2+x}$.

Solution

Because $\lim\limits_{x \to \infty} \dfrac{1}{2+x} = 0$, therefore $f(x)$ has a horizontal asymptote of $y = 0$.

Meanwhile, because $f(x)$ is undefined at $x = -2$, it has a vertical asymptote of $x = -2$ because $\lim\limits_{x \to -2} \dfrac{1}{2+x}$ will be infinite. This can also be verified by finding the horizontal asymptote of its inverse function.

$$x = \frac{1}{y} - 2 \implies \lim_{y \to \infty} \left(\frac{1}{y} - 2\right) = -2$$

Done.

In physics, the Lorentz factor $\gamma(v)$ is a function of speed v which is defined as

$$\gamma(v) = \frac{1}{\sqrt{1 - \frac{v^2}{c^2}}} \tag{2.6}$$

where c is a constant denoting the speed of light in a vacuum.

Example 2.6.2

Find vertical asymptotes of the Lorenz factor function *(2.6)*.

Solution

The vertical asymptotes can be determined by finding the value of v so that $\gamma(v)$ becomes infinity or, equivalently, the denominator

Chapter 2: Limit

of *(2.6)* equals 0.

$$\sqrt{1-\frac{v^2}{c^2}} \implies v = c \qquad (\because v \geq 0)$$

Hence, this function has a vertical asymptote at $v = c$. This can also be observed from its function graph.

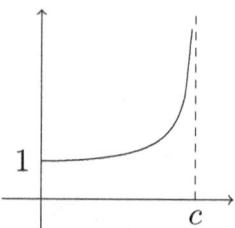

Done.

An interpretation of this result from physics perspective is that an object will require an infinite amount of energy in order to travel at the speed of light. In other words, no object can travel at a speed greater than speed of the light.

2.6.2 Sum Infinite Series

An infinite series is a sequence containing an unlimited number of terms. Infinite series is an important topic in calculus. The whole *Chapter 5* is dedicated to this topic. The first step to investigate an infinite series is to study its sum $S = \sum_{n=1}^{\infty} a_n$. Calculating such a sum requires limit computation. This is already shown in *Example 2.2.1* on *page 6*.

Here is another example.

Example 2.6.3

Show that

$$\frac{1}{1-x} = 1 + x + x^2 + x^3 + x^4 + \cdots \qquad (|x| < 1) \qquad (2.7)$$

Solution

Let
$$S_n = 1 + x + x^2 + x^3 + \cdots + x^n = \frac{1 - x^{n+1}}{1 - x}$$

Then because
$$|x| < 1 \implies x^{n+1} \to 0$$

we find
$$\lim_{n \to \infty} S_n = \lim_{n \to \infty} \frac{1 - x^{n+1}}{1 - x} = \frac{1}{1 - x}$$

Done.

2.6.3 Compute Area Under A Curve

The conclusion of *Example 2.2.2* on *page 7* shows that

$$\lim_{n \to \infty} \sum_{k=1}^{n} \frac{1}{n} \left(\frac{k}{n}\right)^2 = \frac{1}{3} \qquad (2.8)$$

This result is essentially the area below the curve $y = x^2$ over the interval $[0, 1]$. As it will be seen later, calculating the area below a curve is one of the basic tasks in integral.

Chapter 2: Limit

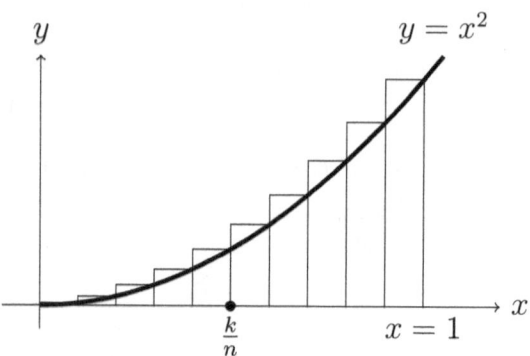

In order to understand that *(2.8)* represents the area below $y = x^2$, let's consider the diagram above. The said area can be approximated by adding up the areas of n equal width rectangles. When n approaches infinity, the sum will give the desired result.

The x-coordinates of the right-bottom corner of these rectangles are $\frac{k}{n}$ where $k = 1$ to n. Accordingly, the heights will be $\left(\frac{k}{n}\right)^2$ because their corresponding right-top corners locate on the curve $y = x^2$. Thus, the area of each rectangle will be $\frac{1}{n}\left(\frac{k}{n}\right)^2$ and the sum will be

$$\sum_{k=1}^{n} \frac{1}{n}\left(\frac{k}{n}\right)^2$$

Taking n to infinity gives *(2.8)*.

2.6.4 Continuously Compounded Interest

In real life, interest on bank deposits is paid periodically. The total balance after t years will be (assuming no withdrawal during this time period)

$$P\left(1 + \frac{r}{n}\right)^{nt} \qquad (2.9)$$

where P is the initial deposit principal, r is annual interest rate and n is the number of paying periods per year. For example, if interest is paid annually, then n will be 1, If interest is paid monthly, then $n = 12$. If interested is paid quarterly, then $n = 4$.

Please note that, strictly speaking, if $n > 1$, then the equivalent period interest rate will be different from $\frac{r}{n}$ in order to accommodate the compounding effect. However, for every n, there exists an equivalent r' that the value $\frac{r'}{n}$ is the true period interest rate. Therefore, the result will be still in the form of (2.9). For simplicity, let's continue using this expression[2].

In the world of professional finance, interest is often assumed to be paid instantly. This means that accrued interest will immediately be added to the principal and start earning additional interest. This is called continuously compounding. Mathematically, it is equivalent to making n in (2.9) infinitely large. Accordingly, the balance after t years becomes

$$\lim_{n \to \infty} P \left(1 + \frac{r}{n}\right)^{nt} = P \left(\lim_{n \to \infty} \left(\left(1 + \frac{r}{n}\right)^{\frac{n}{r}}\right)^{rt}\right) \quad (2.10)$$

$$= P \left(\lim_{n \to \infty} \left(1 + \frac{r}{n}\right)^{\frac{n}{r}}\right)^{rt}$$

$$= \boxed{P \times e^{rt}}$$

where e is the Euler's number mentioned at the end of *Section 2.4*.

The benefit of using continuously compounded interest comes from many useful properties associated with the function e^x. These features are highly desirable in quantitative finance.

---

[2]In fact, most real life applications such as mortgage uses this simplified convention to convert an annual rate to a monthly or daily rate.

Chapter 2: Limit

2.7 Practice

Practice 1

Can the ε-δ definition of limit on page 4 be modified as: for any arbitrary positive number ε, there always exists a number δ such that there are infinitely many $x \in [c-\delta,\ c+\delta]$ satisfying $|f(x) - L| < \varepsilon$?

Practice 2

Evaluate
$$\lim_{n \to \infty} \frac{2n^3 + 99n^2 + 1}{3n^3 + n^2 + 3}$$

Practice 3

Use the Sandwich theorem to evaluate
$$\lim_{x \to \infty} \frac{\sin x}{x}$$

Practice 4

Compute the value of
$$\lim_{n \to \infty} \left(\sqrt{n+1} - \sqrt{n}\right)$$

Practice 5

Compute
$$\lim_{x \to 4} \frac{3 - \sqrt{x+5}}{x - 4}$$

Practice 6

Find the value of $\lim\limits_{n\to\infty} \sin^2\left(\pi\sqrt{n^2+n}\right)$.

(China)

Practice 7

Use the trigonometric difference to product identity below

$$\sin\alpha - \sin\beta = 2\sin\frac{\alpha-\beta}{2}\cos\frac{\alpha+\beta}{2}$$

and the conclusion of *Example 2.3.1* on *page 8* to evaluate

$$\lim_{\Delta x\to 0} \frac{\sin(x+\Delta x) - \sin x}{\Delta x}$$

Practice 8

Recall

$$\sum_{k=1}^{n} k^3 = \left(\frac{n(n+1)}{2}\right)^2$$

Find the area below the curve $y = x^3$ between $x = 0$ and 1.

Practice 9

Use the sandwich theorem to find the value of

$$\lim_{n\to\infty} \sum_{k=1}^{n} \frac{n+k}{n^2+k}$$

Chapter 2: Limit

Practice 10

Apply the bounded monotonic function method to show that $e = \lim\limits_{n\to\infty} \left(1 + \dfrac{1}{n}\right)^n$ exists.

Practice 11

Show that
$$\lim_{x\to 0} \frac{e^x - 1}{x} = 1$$

Chapter 3

Derivative

3.1 Derivative Defined

The derivative of a function $f(x)$ is written and defined as

$$\frac{\mathrm{d}f(x)}{\mathrm{d}x} = \lim_{\Delta x \to 0} \frac{\Delta f(x)}{\Delta x} = \lim_{\Delta x \to 0} \frac{f(x + \Delta x) - f(x)}{\Delta x} \qquad (3.1)$$

It represents the ratio of the change in function value against the change in variable when the latter approaches 0.

When there is no confusion, derivative can also be written as

$$\frac{\mathrm{d}f(x)}{\mathrm{d}x} = \frac{\mathrm{d}}{\mathrm{d}x}f(x) = f'(x) = \frac{\mathrm{d}f}{\mathrm{d}x} = f'$$

From the geometry perspective, $f'(x)$ represents the slope of the tangent line to the curve $y = f(x)$. To see this, let $P(x, y)$ be a fixed point on this curve and $Q(x + \Delta x, y + \Delta y)$ be a movable point within the neighborhood of P. Then, the secant $\overline{PQ}$ will approach and eventually become the tangent line at P when Q moves towards and eventually overlaps with P.

This is shown in the diagram on the next page.

Chapter 3: Derivative

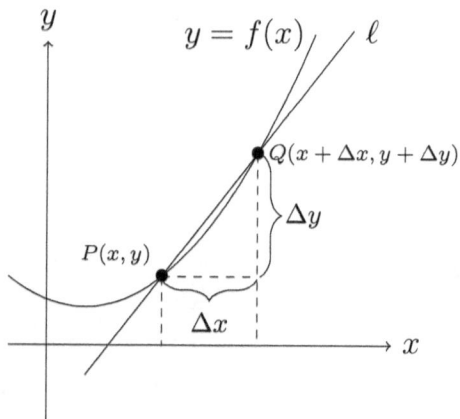

In the diagram above, it is clear that the slope of the secant $\overline{PQ}$ equals $\Delta y/\Delta x$. When Q approaches P, Δx will approach 0 and so will $\Delta y = f(x+\Delta x) - f(x)$ be. Consequently, we find the slope of the tangent line at P equals

$$\lim_{\Delta x \to 0} \frac{\Delta y}{\Delta x} = \lim_{\Delta x \to 0} \frac{f(x+\Delta x) - f(x)}{\Delta x}$$

Because a limit may or may not exist, therefore a derivative may or may not exist as a result.

Example 3.1.1

Explain that the derivative of a constant is 0.

Solution

Let C be a constant. Then the graph of $y = C$ is a horizontal line whose tangent line is itself. Thus, the slope of its tangent line is 0 everywhere. Then, the conclusion follows.

Done.

This conclusion can also be verified using the derivative's definition *(3.1)*. When $f(x)$ is a constant, the numerator, $f(x+\Delta x) - f(x)$, always equals 0. Thus, the derivative will always equal 0.

The derivative of a function is still a function. Therefore, it is possible to take derivative of $f'(x)$ again. The result is called the second order derivative, or second derivative in short, and can be written as

$$\frac{\mathrm{d}^2 f(x)}{\mathrm{d}x^2} = \frac{\mathrm{d}^2}{\mathrm{d}x^2} f(x) = \frac{\mathrm{d}^2 f}{\mathrm{d}x^2} = f''(x) = f''$$

Accordingly, $f'(x)$ can be explicitly called the first derivative. The derivative of the second derivative is called the third derivative. This process can be repeated to obtain the nth derivative if it exists.

3.2 Differentiability vs Continuity

Taking derivative of $f(x)$ is also called differentiating $f(x)$. If the derivative of $f(x)$ exists at point $x = x_0$, then $f(x)$ is said to be differentiable at this point. If the second derivative also exists at this point, then $f(x)$ is said to be twice differentiable at $x = x_0$. If $f(x)$ is differentiable everywhere in its entire domain, then $f(x)$ is called a differentiable function.

For $f(x)$ to be differentiable at point $x = x_0$, it must be continuous at this point. However, its inverse statement is not true. One counter example where a continuous curve is not differentiable everywhere is $y = |x|$.

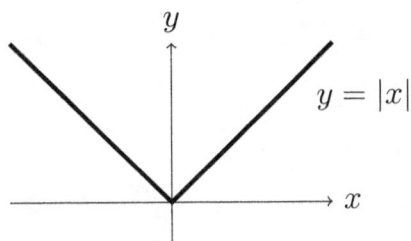

Chapter 3: Derivative

This function is obvious continuous at $x = 0$. But it is not differentiable at this point because

$$\lim_{\Delta x \to 0^+} \frac{|0 + \Delta x| - |0|}{\Delta x} = \lim_{\Delta x \to 0^+} \frac{|\Delta x|}{\Delta x} = \lim_{\Delta x \to 0^+} \frac{\Delta x}{\Delta x} = 1$$

and

$$\lim_{\Delta x \to 0^-} \frac{|0 + \Delta x| - |0|}{\Delta x} = \lim_{\Delta x \to 0^-} \frac{|\Delta x|}{\Delta x} = \lim_{\Delta x \to 0^-} \frac{-\Delta x}{\Delta x} = -1$$

The left and right limits do not agree at this point. Therefore, the limit does not exist, neither does the derivative.

In fact, there exist functions which are continuous everywhere but differentiable nowhere. One such example is the Weierstrass function.

3.3 Compute Derivative (I)

Applying derivative's definition is one method to compute a derivative. Below is such an example:

Example 3.3.1

Compute the derivative of $f(x) = x^n$.

Solution

Applying *(3.1)* and binomial expansion gives

$$\frac{\mathrm{d}f(x)}{\mathrm{d}x} = \lim_{\Delta x \to 0} \frac{f(x + \Delta x) - f(x)}{\Delta x}$$

$$= \lim_{\Delta x \to 0} \frac{(x + \Delta x)^n - x^n}{\Delta x}$$

$$= \lim_{\Delta x \to 0} \frac{\left(x^n + nx^{n-1}\Delta x + \binom{n}{2}x^{n-1}(\Delta x)^2 + \cdots\right) - x^n}{\Delta x}$$

$$= \lim_{\Delta x \to 0} \left(nx^{n-1} + \binom{n}{2} x^{n-2} \Delta x + \cdots \right)$$
$$= nx^{n-1}$$

Therefore, we conclude

$$\boxed{\frac{\mathrm{d}}{\mathrm{d}x} x^n = nx^{n-1}} \quad (3.2)$$

Done.

This result, known as the power rule, is one of the basic derivative rules. It is worth noting that *(3.2)* still holds even if n is not a positive integer. When n is not a positive integer, the proof above can be amended by using the generalized binomial expansion theorem[1]. The conclusion will stay the same. As applications of *(3.2)*, setting $n = -1$ and $\frac{1}{2}$, respectively, gives

$$\frac{\mathrm{d}}{\mathrm{d}x} \frac{1}{x} = \frac{\mathrm{d}}{\mathrm{d}x} x^{-1} = (-1)x^{-2} = -\frac{1}{x^2} \quad (3.3)$$

$$\frac{\mathrm{d}}{\mathrm{d}x} \sqrt{x} = \frac{\mathrm{d}}{\mathrm{d}x} x^{\frac{1}{2}} = \frac{1}{2} x^{-\frac{1}{2}} = \frac{1}{2\sqrt{x}} \quad (3.4)$$

Conclusions such as *(3.2)*, *(3.3)* and *(3.4)* need to be memorized because they are basic building blocks of differentiating more complex functions.

Example 3.3.2

Given the function $f(x) = x^2$, what is the slope of the tangent line at the point $(1, 1)$?

The slope of the tangent line is the derivative.

---

[1]The extended binomial expansion theorem is discussed in the book *More on Counting*.

Chapter 3: Derivative

Solution

Setting $n = 2$ in *(3.2)* gives $f'(x) = 2x$. Then, setting $x = 1$ in the derivative leads to the final result $\boxed{2}$.

This result can be verified by noting that the tangent line equation for $y = x^2$ at point (x_0, y_0) is

$$\frac{y + y_0}{2} = xx_0 \implies y = 2x_0 x - y_0$$

Hence, the slope at $x_0 = 1$ equals $2 \times 1 = 2$.

Done.

3.4 Derivative's Properties

Derivative is additive. This means that

$$\frac{d}{dx}(f(x) \pm g(x)) = \frac{d}{dx}f(x) \pm \frac{d}{dx}g(x) \qquad (3.5)$$

This property, also known as the sum and difference rule, can be derived by applying *(3.1)* directly:

$$\begin{aligned}
&\frac{d}{dx}(f(x) \pm g(x)) \\
&= \lim_{\Delta x \to 0} \frac{(f(x + \Delta x) \pm g(x + \Delta x)) - (f(x) \pm g(x))}{\Delta x} \\
&= \lim_{\Delta x \to 0} \frac{f(x + \Delta x) - f(x)}{\Delta x} \pm \lim_{\Delta x \to 0} \frac{g(x + \Delta x) - g(x)}{\Delta x} \\
&= \frac{d}{dx}f(x) \pm \frac{d}{dx}g(x)
\end{aligned}$$

Chapter 3: Derivative

> Let C be a constant, then
>
> $$\frac{\mathrm{d}}{\mathrm{d}x}(Cf(x)) = C\left(\frac{\mathrm{d}}{\mathrm{d}x}f(x)\right) \qquad (3.6)$$

This property, sometimes noted as the constant multiplication rule, can also be proved directly using (3.1). However, it is important to note that if C is a not a constant, then (3.6) does not hold. In other words, the derivative of a product of two functions does not equal the product of two individual derivatives.

$$\frac{\mathrm{d}}{\mathrm{d}x}(f(x)g(x)) \neq \left(\frac{\mathrm{d}}{\mathrm{d}x}f(x)\right)\left(\frac{\mathrm{d}}{\mathrm{d}x}g(x)\right)$$

This is because

$$\frac{\mathrm{d}}{\mathrm{d}x}(f(x)g(x)) = \lim_{\Delta x \to 0} \frac{f(x+\Delta x)g(x+\Delta x) - f(x)g(x)}{\Delta x}$$

which usually does not equal

$$\left(\frac{\mathrm{d}}{\mathrm{d}x}f(x)\right)\left(\frac{\mathrm{d}}{\mathrm{d}x}g(x)\right)$$
$$= \left(\lim_{\Delta x \to 0} \frac{f(x+\Delta x) - f(x)}{\Delta x}\right)\left(\lim_{\Delta x \to 0} \frac{g(x+\Delta x) - g(x)}{\Delta x}\right)$$
$$= \lim_{\Delta x \to 0} \left(\frac{f(x+\Delta x) - f(x)}{\Delta x} \cdot \frac{g(x+\Delta x) - g(x)}{\Delta x}\right)$$

Instead, derivative's product rule states

> **Theorem 3.4.1 The Product Rule**
>
> $$\frac{\mathrm{d}\,(u(x)v(x))}{\mathrm{d}x} = u(x)\frac{\mathrm{d}v(x)}{\mathrm{d}x} + v(x)\frac{\mathrm{d}u(x)}{\mathrm{d}x} \qquad (3.7)$$

Equation 3.7 can also be written more concisely as

$$(uv)' = uv' + vu' \qquad (3.8)$$

The constant coefficient rule *(3.6)* is a special case of the product rule because the derivative of a constant equals 0.

The product rule can also be proved using definition *(3.1)* and polynomial transformation. The proof will be left as a practice.

3.5 Derivative of e^x

The function e^x is one of the most remarkable functions in calculus due to the following property:

$$e^x = (e^x)' = (e^x)'' = (e^x)''' = \cdots \qquad (3.9)$$

In other words, the function e^x is infinitely differentiable and its derivative is always itself. To show this, it is sufficient to prove $(e^x)' = e^x$. If this can be proved, then repeatedly differentiating both sides will lead to *(3.9)*.

Example 3.5.1

Prove $(e^x)' = e^x$.

Proof

By derivative's definition:

$$\frac{de^x}{dx} = \lim_{\Delta x \to 0} \frac{e^{x+\Delta x} - e^x}{\Delta x} = \lim_{\Delta x \to 0} \frac{e^x\left(e^{\Delta x} - 1\right)}{\Delta x} = e^x \lim_{\Delta x \to 0} \frac{e^{\Delta x} - 1}{\Delta x}$$

From previous chapter's practice, we have $\lim_{x \to 0} \frac{e^x - 1}{x} = 1$. Setting this to the above relation leads to the desired result.

$$QED$$

3.6 Notation Demystification

Derivative notations often appear to be abstract and confusing to beginners. As a result, many students feel it difficult to transform and manipulate derivative expressions. This section aims to help students become more comfortable with derivative operations and notations.

First, $\mathrm{d}x$ can be viewed as Δx when the latter approaches 0, i.e. $\mathrm{d}x = \lim_{\Delta x \to 0} \Delta x$. This understanding is consistent with derivative's definition because when $\Delta x \to 0$, so will Δy.

$$\frac{\mathrm{d}y}{\mathrm{d}x} = \lim_{\Delta x \to 0} \frac{\Delta y}{\Delta x}$$

Substituting $\mathrm{d}x$ using Δx is often an effective way to help beginners to understand some abstract complex expressions. This tip will be used later in this book.

Next, differentiation is just a regular function which can be used in the same way as other functions. For example, given two equal variables x and y, it is possible to apply the same function on these two variables and still maintain the equality, such as:

$$x = y \implies x^2 = y^2, \quad \sin x = \sin y, \quad a^x = a^y, \quad \ldots$$

In a similar way, it is also valid to differentiate both side of an equation. For example,

$$x = y \implies \mathrm{d}x = \mathrm{d}y \tag{3.10}$$

Additionally, $\mathrm{d}x$ can be used as a single operating block in the same way as $\sin x$ etc. For instance, the following transformation is frequently used.

$$\frac{\mathrm{d}y}{\mathrm{d}x} = u \implies \mathrm{d}y = u\mathrm{d}x \tag{3.11}$$

Chapter 3: Derivative

This is similar to the following assertion:

$$\frac{\sin y}{\sin x} = 2 \implies \sin y = 2\sin x$$

Another way to understand *(3.11)* is to substitute dx and dy with Δx and Δy as discussed earlier.

$$\frac{\Delta y}{\Delta x} = u \implies \Delta y = u\Delta x$$

This expression may be easier to understand.

Finally, it is worth emphasizing that x and dx are two different types of quantities, or have different "units". This is similar to the fact that distance is different from speed even though they numerically may look the same. Therefore, expressions such as $x = dx$ or $y = dx$ will not hold. Instead, expressions such as $dy = dx$, $ydy = udx$, and so on are semantically correct.

3.7 Implicit Derivative

An application of *(3.10)* is to compute implicit derivative. Sometimes, it is more convenient to describe a relation using an implicit form of $f(x, y) = 0$, rather than in an explicit form of $y = f(x)$. In such cases, it is often possible to find the derivative dy / dx directly without first converting the implicit relation to an explicit form.

Let's consider an example:

Example 3.7.1

Find the slope at point (x_0, y_0) on the circle $x^2 + y^2 = R^2$ where R is a constant.

Solution

Recall that the slope at point (x_0, y_0) equals the value of its derivative at that point. In order to compute its derivative $\frac{dy}{dx}$, let's first take derivatives on both sides of the given equation:

$$\begin{aligned} d\left(x^2 + y^2\right) &= dR^2 \\ dx^2 + dy^2 &= 0 & (\because R \text{ is a constant}) \\ 2x\,dx + 2y\,dy &= 0 & (\because \text{Equation 3.2}) \\ \therefore \quad \frac{dy}{dx} &= -\frac{x}{y} \end{aligned}$$

Therefore, the slope at point (x_0, y_0) is $-\frac{x_0}{y_0}$.

Done.

This result can also be obtained by first converting the relation to an explicit form $y = \pm\sqrt{R^2 - x^2}$. This alternative solution will be left as a practice.

3.8 Differentiate Parametric Function

Differentiating a parametric function is also a good example of utilizing *(3.10)*. Let a parametric function be defined as

$$\begin{cases} x = g(t) \\ y = h(t) \end{cases}$$

Then

$$\frac{dy}{dx} = \frac{dy/dt}{dx/dt} \quad \left(\frac{dx}{dt} \neq 0\right) \tag{3.12}$$

The denominator is obtained by taking derivative of x's parametric function, and the numerator is the derivative of y's parametric function.

Let's review an example.

Chapter 3: Derivative

Example 3.8.1

Find the parametric equation of the parabola $y = x^2$ and compute its derivative.

Solution

The desired parametric equation is

$$\begin{cases} x = t \\ y = t^2 \end{cases}$$

Therefore,

$$\frac{dy}{dx} = \frac{dy / dt}{dx / dt} = \frac{2t}{1} = \boxed{2t}$$

Done.

In this particular case, because $x = t$, therefore the result $2t$ is equivalent to $2x$. This is the same as if taking derivative directly:

$$y = x^2 \implies \frac{dy}{dx} = 2x$$

3.9 Compute Derivative (II)

Most derivatives are not computed using definitions as shown in *Section 3.3*. Instead, computations are usually carried out using various basic formulas and rules. One of them is to utilize *(3.10)* and *(3.11)*. They can be used to derive new results based on known ones. Here is an example.

Example 3.9.1

Given $(e^x)' = e^x$, find the derivative of $y = \ln x$.

Solution

Let's express x in terms of y and then differentiate both sides:
$$y = \ln x \implies x = e^y \implies \mathrm{d}x = \mathrm{d}\left(e^y\right)$$

Meanwhile, we have
$$\frac{\mathrm{d}\left(e^y\right)}{\mathrm{d}y} = e^y \implies \mathrm{d}\left(e^y\right) = e^y \mathrm{d}y$$

Setting this to the previous relation yields
$$\mathrm{d}x = \mathrm{d}\left(e^y\right) \implies \mathrm{d}x = e^y \mathrm{d}y \implies \frac{\mathrm{d}y}{\mathrm{d}x} = \frac{1}{e^y}$$

Note $e^y = x$. Therefore, the desired result is
$$\frac{\mathrm{d}y}{\mathrm{d}x} = \boxed{\frac{\mathrm{d}}{\mathrm{d}x}\ln x = \frac{1}{x}} \tag{3.13}$$

Done.

3.10 Inverse Function Rule

If the derivative of a function $f(x)$ is known and $f(x)$ has a well-defined inverse $g(x) = f^{-1}(x)$, then

> **Theorem 3.10.1 The Inverse Function Rule**
>
> If $g(x)$ is the inverse of $f(x)$, then
> $$g'(x) = \frac{1}{f'(g(x))} \tag{3.14}$$

Equation 3.14 can be understood intuitively by using the following elementary geometric properties. Let function $g(x)$ be the inverse of function $f(x)$, then

i) the plots of these two function are symmetric with respect to the line $y = x$.

ii) the mirroring point of (x, y) on curve $g(x)$ is the point (y, x) on curve $f(x)$.

iii) the slope of the tangent line at (x, y) on $g(x)$ is the multiplicative inverse of the slope of the tangent line at (y, x) on $f(x)$. In another word, the product of these two slopes equals 1.

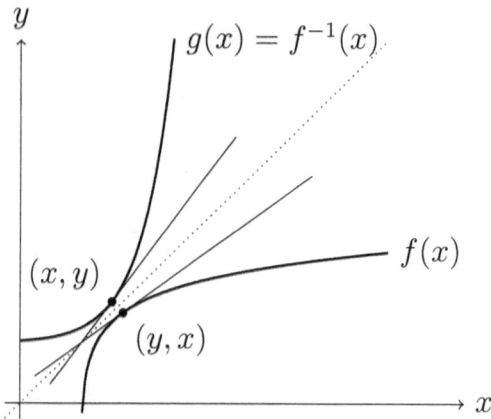

In order to determine the derivative of $g(x)$, let's take an arbitrary point (x, y) where $y = g(x)$. Accordingly, its mirroring point on $f(x)$ is (y, x). Because the slopes of the tangents at (x, y) and (y, x) are $g'(x)$ and $f'(y)$, respectively, therefore we find

$$g'(x)f'(y) = 1 \implies g'(x) = \frac{1}{f'(y)} = \frac{1}{f'(g(x))}$$

Let's revisit the previous example to see how this inverse function rule can be used.

Example 3.10.1

Given $(e^x)' = e^x$, find the derivative of $\ln x$.

Chapter 3: Derivative

Solution

Function $\ln x$ is the inverse of e^x. Let $g(x) = \ln x$, $f(x) = e^x$, and $f'(x) = e^x$. Then

$$g'(x) = \frac{1}{f'(g(x))} = \frac{1}{e^{g(x)}} = \frac{1}{e^{\ln x}} = \frac{1}{x}$$

This agrees with *(3.13)* on *page 37.*

Done.

Other applications of this inverse function rule include differentiating inverse trigonometric functions. This will be shown in the next section.

3.11 Derivatives of Trigonometric Functions

In the previous chapter's practice, we have already shown that

$$\lim_{\Delta x \to 0} \frac{\sin(x + \Delta x) - \sin x}{\Delta x} = \cos x$$

This essentially means that

$$\boxed{(\sin x)' = \cos x} \qquad (3.15)$$

Using a similar technique, we can find the derivative of $\cos x$:

$$\boxed{(\cos x)' = -\sin x} \qquad (3.16)$$

Deriving *(3.16)* will need the following identity:

$$\cos \alpha - \cos \beta = -2 \sin \frac{\alpha + \beta}{2} \cos \frac{\alpha - \beta}{2}$$

Chapter 3: Derivative

Computing the derivative of the tangent function $\tan x$ will require the chain rule or the quotient rule because it is the quotient of $\sin x$ and $\cos x$. They will be discussed later. Here, let's find inverse trigonometric functions' derivatives.

Example 3.11.1

Find the derivative of $\arcsin x$.

Solution

Let $f(x) = \sin x$, $g(x) = f^{-1}(x) = \arcsin x$. Then $f'(x) = \cos x$,

$$g'(x) = \frac{1}{f'(g(x))} = \frac{1}{\cos(\arcsin x)}$$

Let $u = \arcsin x$, then

$$x = \sin u \implies \cos u = \sqrt{1-x^2} \implies \cos(\arcsin x) = \sqrt{1-x^2}$$

$$\therefore \quad g'(x) = \boxed{(\arcsin x)' = \frac{1}{\sqrt{1-x^2}}} \qquad (3.17)$$

Done.

Similarly, it can be shown that

$$\boxed{(\arccos x)' = -\frac{1}{\sqrt{1-x^2}}} \qquad (3.18)$$

3.12 The Chain Rule

The chain rule deals with derivative of a composite function, i.e. a function of function. An intuitive way to understand the chain rule is to use *Equation 3.11* on *page 33*.

Chapter 3: Derivative

Theorem 3.12.1 The Chain Rule

The derivative of $f(g(x))$ can be computed as

$$\frac{df}{dx} = \frac{df}{dg} \cdot \frac{dg}{dx} \qquad (3.19)$$

Equation 3.19 means that two steps are required when differentiating a composite function. The first step is to treat the inside function as a whole and compute the derivative against it. The second step is to differentiate the function inside. Then, the product of these two derivatives is the final result.

Example 3.12.1

Compute the derivative of $f(x) = (x+1)^2$.

Solution

Let $g = x + 1$, then $f = g^2$. Therefore,

$$\frac{df}{dx} = \frac{df}{dg} \cdot \frac{dg}{dx} = (2g)(1+0) = (2(x+1))(1) = \boxed{2(x+1)}$$

The chain rule can also be applied without explicit substitution:

$$\frac{d(x+1)^2}{dx} = \frac{d(x+1)^2}{d(x+1)} \cdot \frac{d(x+1)}{dx} = 2(x+1) \cdot 1 = \boxed{2(x+1)}$$

Students are encouraged to use the second form when they become comfortable about the chain rule. Familiar with this style will lay a necessary foundation for integrating using the substitution method. This topic will be discussed in the next chapter.

Done.

The result of the previous example can also be verified by first

Chapter 3: Derivative

expanding $f(x)$ and then taking derivative:

$$f'(x) = (x^2 + 2x + 1)' = 2x + 2 + 0 = 2(x+1)$$

Let's consider another example.

Example 3.12.2

Use two different approaches to compute the derivative of $f(x) = \sin(2x)$.

Solution by the chain rule

In order to use the chain rule, we need to treat $2x$ as a whole,

$$\frac{\mathrm{d}\sin(2x)}{\mathrm{d}x} = \frac{\mathrm{d}\sin(2x)}{\mathrm{d}(2x)} \cdot \frac{\mathrm{d}(2x)}{\mathrm{d}x} = (\cos 2x) \cdot 2 = \boxed{2\cos 2x}$$

Done.

Solution by the product rule

The double angle formula says $\sin(2x) = 2\sin x \cos x$. Therefore

$$\begin{aligned}
(\sin(2x))' &= (2\sin x \cos x)' \\
&= 2(\sin x \cos x)' \\
&= 2\left(\sin x (\cos x)' + (\sin x)' \cos x\right) \\
&= 2\left(\sin x(-\sin x) + \cos x \cos x\right) \\
&= 2(\cos^2 x - \sin^2 x) \\
&= \boxed{2\cos(2x)}
\end{aligned}$$

Done.

Example 3.12.3

Compute $(\cos x)'$ using the chain rule.

Solution

This can be achieved using the identity: $\cos x = \sin\left(\frac{\pi}{2} - x\right)$:

$$\frac{d\cos x}{dx} = \frac{d\sin\left(\frac{\pi}{2} - x\right)}{dx}$$

$$= \frac{d\sin\left(\frac{\pi}{2} - x\right)}{d\left(\frac{\pi}{2} - x\right)} \cdot \frac{d\left(\frac{\pi}{2} - x\right)}{dx}$$

$$= \cos\left(\frac{\pi}{2} - x\right) \cdot (-1)$$

$$= -\sin x$$

Done.

3.13 The Quotient Rule

The quotient of two functions $f(x)/g(x)$ can be viewed as a product of $f(x)$ and $(g(x))^{-1}$. The latter can be computed using the chain rule. Therefore, the derivative of a quotient can be calculated by combining the product rule and the chain rule:

$$\left(\frac{f}{g}\right)' = \left(f \cdot \frac{1}{g}\right)' = f\left(\frac{1}{g}\right)' + f' \cdot \frac{1}{g} = f\left(-\frac{g'}{g^2}\right) + \frac{f'}{g}$$

Simplifying the last term gives the quotient rule as

$$\left(\frac{f}{g}\right)' = \frac{-fg' + f'g}{g^2} \qquad (3.20)$$

Chapter 3: Derivative

Let's consider an example.

Example 3.13.1

Compute the derivative of the tangent function

Proof

Applying the quotient rule gives

$$(\tan x)' = \left(\frac{\sin x}{\cos x}\right)' = \frac{-\sin x (\cos x)' + (\sin x)' \cos x}{\cos^2 x}$$

$$= \frac{-\sin x(-\sin x) + \cos x \cos x}{\cos^2 x}$$

$$= \frac{1}{\cos^2 x}$$

Therefore,

$$\boxed{\frac{d}{dx} \tan x = \frac{1}{\cos^2 x} = \sec^2 x} \qquad (3.21)$$

QED

The derivative of $(\cot x)'$ can be obtained in a similar way.

$$\boxed{\frac{d}{dx} \cot x = -\frac{1}{\sin^2 x} = -\csc^2 x} \qquad (3.22)$$

3.14 Table of Common Derivatives

Computing derivatives usually can be carried out in a systematical way: determine the derivatives of some basic functions and then applying applicable rules (e.g. the chain rule, the product rule, etc) as necessary to obtain the final result. Hence, it is imperative to memorize the relations listed in the table below.

Constant Rule	$C' = 0$	See *Example 3.1.1*.
Power Rule	$(x^n)' = nx^{n-1}$	See *(3.2)* on *page 29*.
Exponent Rule	$(e^x)' = e^x$	See *(3.9)* on *page 32*.
	$(a^x)' = a^x \ln a$	See practice.
Logarithm Rule	$(\ln x)' = \frac{1}{x}$	See *(3.13)* on *page 37*.
	$(\log_a x)' = \frac{1}{x \ln a}$	See practice.
Trigonometry Rule	$(\sin x)' = \cos x$	See *(3.15)* on *page 39*
	$(\cos x)' = -\sin x$	See *(3.16)* on *page 39*.
	$(\tan x)' = \sec^2 x$	See *(3.21)* on *page 44*.
	$(\cot x)' = -\csc^2 x$	See *(3.22)* on *page 44*.
	$(\arcsin x)' = \frac{1}{\sqrt{1-x^2}}$	See *(3.17)* on *page 40*.
	$(\arccos x) = -\frac{1}{\sqrt{1-x^2}}$	See *(3.18)* on *page 40*.
	$(\arctan x)' = \frac{1}{1+x^2}$	See practice.

3.15 Additional Techniques

3.15.1 Differentiate x^x

When the variable x appears at both the base and the exponent, a proper transformation is often required. Let's review an example.

Chapter 3: Derivative

Example 3.15.1

Find the derivative of x^x.

Solution

Using $x^x = e^{x \ln x}$ and applying the chain rule give

$$f(x) = e^{x \ln x} \implies f'(x) = e^{x \ln x}\left(\ln x + x \cdot \frac{1}{x}\right) = \boxed{x^x(\ln x + 1)}$$

This problem can also be solved using the technique employed in *Example 3.9.1* on *page 36*.

Let $y = x^x$, then $\ln y = x \ln x$. Differentiating both sides yields

$$\frac{1}{y}\,dy = (1 + \ln x)\,dx \implies \frac{dy}{dx} = y(1 + \ln x) = \boxed{x^x(1 + \ln x)}$$

Done.

Transforming a^x to $e^{x \ln a}$ is a useful technique.

3.15.2 Product Rule For Higher Order

The product rule can be extended to compute the derivative of order higher than 1. Let's first investigate the 2nd and the 3rd orders and then generalize the result by observing the pattern.

Because $(uv)' = u'v + uv'$, therefore differentiating both sides again gives

$$(uv)'' = (u'v)' + (uv')' = u''v + u'v' + u'v' + uv''$$

Merging like terms yields

$$(uv)'' = u''v + 2u'v' + uv'' \tag{3.23}$$

Differentiate again on *(3.23)* and merging like terms will yield

$$(uv)''' = u'''v + 3u''v' + 3u'v'' + uv''' \tag{3.24}$$

Observing the patterns exhibited in *(3.23)* and *(3.24)* reveals their similarity with the binomial expansion. By mathematical induction, it can be shown that

$$(uv)^{(n)} = \sum_{k=0}^{n} \binom{n}{k} u^{(k)} v^{(n-k)} \tag{3.25}$$

This relations can be intuitively explained in a similar way as to explain the binomial expansion. To differentiate (uv) n times means to enumerate all the possible combinations of differentiating u k times and v $(n-k)$ times where $k = 0, 1, \cdots, n$. This is what the right side of *(3.25)* states.

3.16 Concavity

The first derivative represents the slope of the tangent line. Therefore, a positive $f'(x)$ corresponds to an upward $f(x)$ curve, and a negative $f'(x)$ corresponds to a downward $f(x)$ curve. A zero $f'(x)$ indicates a local extreme point (maximum or minimum), or a inflection point (whose exact definition will be given later).

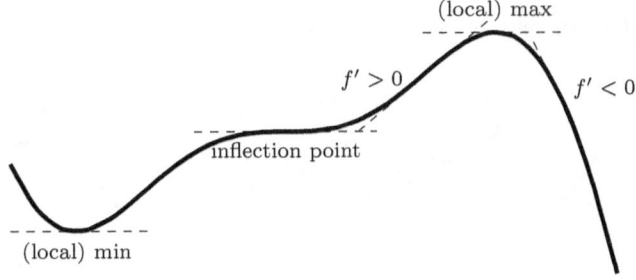

Figure 3.1: first derivative and curve trend

Chapter 3: Derivative

In a similar way, the second derivative describes the trend of the first derivative $f'(x)$. A positive second derivative corresponds to an increasing $f'(x)$. Accordingly, the $f(x)$ curve concaves upwards such as $y = x^2$. And a negative second derivative corresponds to a decreasing $f'(x)$ which translates to a downward concaving $f(x)$ curve such as $y = -x^2$.

When the value of $f'(x)$ changes from positive to negative (or vice versa), $f(x)$ curve will illustrates a complete concavity, as shown below. If the second derivative does not change sign, then $f(x)$ curve will illustrates a "half" concavity. This is shown by the two segments which are before and after the inflection point, respectively, in *Figure 3.1* above. Hence, an *inflection* point is where the curve's concavity changes.

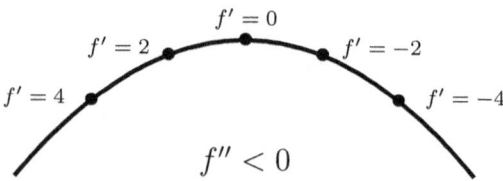

A zero second derivative means no concavity. This corresponds to a linear function $f(x)$ which represents a straight line.

Intuitively, concavity corresponds to the relative positions of the curve and any chord connecting the two points on this curve. If the $f(x)$ curve concaves upward, then the chord will be above the curve. Otherwise, the chord is below the curve.

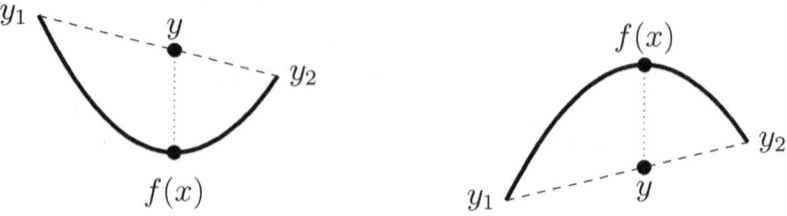

Let the coordinates of the two ending points be (x_1, y_1) and (x_2, y_2). Then, the x-coordinate of any point located between x_1 and x_2 can be written as

$$x = \alpha x_1 + (1 - \alpha) x_2$$

where $\alpha \in [0, 1]$. Then, the y-coordinate of the point on the curve is $f(x)$ and the y-coordinate of the corresponding point on the chord can be obtained by interpolating the y-coordinates of the two ending points, i.e.:

$$y = \alpha f(x_1) + (1 - \alpha) f(x_2)$$

Therefore, for every $0 \leq \alpha \leq 1$ and any pair of two ending points on this curve, a concaving upward curve must satisfies

$$f(\alpha x_1 + (1 - \alpha) x_2) \leq \alpha f(x_1) + (1 - \alpha) f(x_2) \tag{3.26}$$

A concaving downward curve must satisfies

$$f(\alpha x_1 + (1 - \alpha) x_2) \geq \alpha f(x_1) + (1 - \alpha) f(x_2) \tag{3.27}$$

These two relations are closely related to the Jensen's inequality.

3.17 Partial Derivative

Derivatives can also be applied to multi-variable functions. When differentiating a multi-variable function, the variable against which the derivative is computed must be specified. Other non-involving variables should be treated as constants. This is called a *partial derivative*. For example, given a two-variable function

$$f(x, y) = x^2 + xy + 2y^2$$

its partial derivatives with respect to x and y are shown below. Please note the different notations used for derivatives and partial derivatives.

$$\frac{\partial f}{\partial x} = f'_x = 2x + y \quad \text{and} \quad \frac{\partial f}{\partial y} = f'_y = x + 4y$$

Chapter 3: Derivative

A partial derivative can be further differentiated with respect to any of these variables. For example,

$$\frac{\partial^2 f}{\partial x^2} = f'_{xx} = 2, \qquad \frac{\partial^2 f}{\partial x \partial y} = f'_{xy} = 1, \qquad \frac{\partial^2 f}{\partial y \partial x} = f'_{yx} = 1$$

Many real life applications which use calculus involve multi-variable functions and, as a result, require partial derivatives. Such applications may involve one or more multi-variable functions.

Two of the most used partial derivatives are the Hessian matrix and the Jacobian matrix.

Given a single multi-variable function $f(x_1, x_2, \cdots, x_n)$, the Hessian matrix of this function is a $(n \times n)$ matrix of the 2nd partial derivatives which is defined as

$$\mathbb{H} = \begin{pmatrix} \frac{\partial^2 f}{\partial x_1^2} & \frac{\partial^2 f}{\partial x_1 \partial x_2} & \frac{\partial^2 f}{\partial x_1 \partial x_3} & \cdots & \frac{\partial^2 f}{\partial x_1 x_n^2} \\ \frac{\partial^2 f}{\partial x_2 \partial x_1} & \frac{\partial^2 f}{\partial x_2^2} & \frac{\partial^2 f}{\partial x_2 \partial x_3} & \cdots & \frac{\partial^2 f}{\partial x_1 x_n^2} \\ & & \cdots & & \\ \frac{\partial^2 f}{\partial x_n \partial x_1} & \frac{\partial^2 f}{\partial x_n \partial x_2} & \frac{\partial^2 f}{\partial x_n \partial x_3} & \cdots & \frac{\partial^2 f}{\partial x_n^2} \end{pmatrix}$$

Given a system of n multi-variable functions

$$\begin{cases} f_1(x_1, x_2, \cdots, x_m) \\ f_2(x_1, x_2, \cdots, x_m) \\ \cdots \\ f_n(x_1, x_2, \cdots, x_m) \end{cases}$$

the Jacobian matrix of this system is a $(n \times m)$ matrix of partial derivatives which is defined as

$$\mathbb{J} = \begin{pmatrix} \frac{\partial f_1}{\partial x_1} & \frac{\partial f_1}{\partial x_2} & \frac{\partial f_1}{\partial x_3} & \cdots & \frac{\partial f_1}{\partial x_n} \\ \frac{\partial f_2}{\partial x_1} & \frac{\partial f_2}{\partial x_2} & \frac{\partial f_2}{\partial x_3} & \cdots & \frac{\partial f_2}{\partial x_n} \\ & & \cdots & & \\ \frac{\partial f_n}{\partial x_1} & \frac{\partial f_n}{\partial x_2} & \frac{\partial f_n}{\partial x_3} & \cdots & \frac{\partial f_n}{\partial x_n} \end{pmatrix}$$

A Jacobian matrix is often written in a vector format. A system of functions can be written as a vector of functions $\mathbb{F}$:

$$\mathbb{F} = [f_1, \ f_2, \ f_3, \ \cdots, \ f_n]$$

and all the variables can be written in a vector format as well:

$$\mathbb{X} = [x_1, \ x_2, \ x_3, \ \cdots, \ x_m]$$

Then, the Jocobian matrix $\mathbb{J}$ above can also be written as:

$$\mathbb{J} = \frac{\partial \mathbb{F}}{\partial \mathbb{X}} = \left[\frac{\partial \mathbb{F}}{\partial x_1}, \ \frac{\partial \mathbb{F}}{\partial x_2}, \ \frac{\partial \mathbb{F}}{\partial x_3}, \ \cdots, \ \frac{\partial \mathbb{F}}{\partial x_m} \right]$$

Such vector form notation is widely used in practice, such as in machine learning lectures.

3.18 Examples and Applications

3.18.1 Determine Minimum and Maximum

Derivative is a powerful tool to find the local minimal and maximal values of a given function $f(x)$. This is because the first deriva-

Chapter 3: Derivative

tive at these points must be zero or <u>do not exist</u>. An example where the first derivative does not exist at a local minimal is the origin on the curve $y = |x|$.

A point on the curve $f(x)$ where $f'(x)$ equals 0 or does not exist is called a *critical* point. An extreme point must be a critical point. However, a critical point may be not a local extreme. One counter example is the inflection point on *Figure 3.1* on *page 47*.

There are several methods to test whether or not a critical point is a local extreme, and if it is, whether it is a local maximum or minimum.

One method is manual checking. Let c be a critical point on curve $f(x)$. Then, we can find the result by comparing the value of $f(c_k)$ and the values of any points on the left and right of c_k (assuming the chosen point does not go pass a neighboring critical point). Alternatively, it is also possible to check signs of $f'(x)$ instead at the left and right of c because signs of first derivative correspond to the trend of curve.

The third method is to use the second derivative.

- If $f''(c_k) > 0$, then c_k is a local minimum because this point locates in an upward concaving region.

- If $f''(c_k) < 0$, then c_k is a local maximum because it locates in a downward concaving region

- If $f''(c_k) = 0$, then the test is inconclusive. For example, functions x^3 and x^4 both have zero first and second derivative at $x = 0$. However, the point $x = 0$ is an inflection point on curve x^3 but a local minimum on curve x^4.

When $f''(x) = 0$ which causes the previous test to be inconclusive, one solution is to continue checking $f^{(3)}(x)$, $f^{(4)}(x)$, $\cdots$, until a non-zero higher order derivative is found or a higher order derivative ceases to exist. If $f(x)$ is differentiable at least to the nth order where n is even, and

$$f'(c) = f''(c) = f'''(c) = \cdots = f^{(n-1)}(c) = 0, \qquad f^n(c) \neq 0$$

then, if $f^{(n)}(c) > 0$, point $x = c$ be a local minimum. If $f^{(n)}(c) < 0$, point $x = c$ will be a local maximum.

Let's consider an example.

Example 3.18.1

Find all the local maximum and minimum of function

$$f(x) = 2x^3 - 9x^2 + 12x + 8$$

Solution

The first step is to find all critical points of $f(x)$:

$$f'(x) = 6x^2 - 18x + 12 = 0$$

Because $f'(x)$ is defined over the entire $\mathbb{R}$, therefore $f'(x)$ always exists which means that it is sufficient to just consider its zeros.

$$f'(x) = 0 \implies x = 1, \ 2$$

Now, let's determine the sign of $f'(x)$ in the following region:

i) $x < 1 \implies f'(x) > 0$

ii) $1 < x < 2 \implies f'(x) < 0$

iii) $x > 2 \implies f'(x) > 0$

Therefore, we conclude $f(x)$ has two extreme values:

i) A local maximum at $x = 1 \implies f(x) = 13$

ii) A local minimum at $x = 2 \implies f(x) = 12$

Done.

Chapter 3: Derivative

The same result can also be obtained by checking the second derivative. In this case, we have

$$f''(x) = 12x - 18$$

It has a root of $\frac{3}{2}$. Therefore

i) $1 < \frac{3}{2} \implies f''(x) < 0 \implies x = 1$ is a local maximum

ii) $2 > \frac{3}{2} \implies f''(x) > 0 \implies x = 2$ is a local minimum

It is worth emphasizing that a local maximum may not be a global maximum. Even the largest local maximum may not be the global maximum. For instance, in *Example 3.18.1*, the value of $f(x)$ will go up infinitely. But, there is only one local maximum 13.

That being said, global extreme values of a continuous function over a closed interval must exist. This is called the extreme value theorem which is stated as follow:

> **Theorem 3.18.1 Extreme Value Theorem**
>
> If function $f(x)$ is continuous over a closed interval $[a, b]$, then it must have at least one global maximum and at least one global minimum in this interval. These global extremes must be either local extremes or the values at the ending points.

It is critical that the said interval is closed for this conclusion to hold. For example, the function $\frac{1}{x}$ is continuous over $(0, 1]$, but does not have a maximum over this period.

Example 3.18.2

Find the global maximal and minimal value of the function

$$f(x) = 2x^3 - 9x^2 + 12x + 8$$

over the interval $[0, 3]$.

Solution

By the conclusion of *Example 3.18.1* on *page 53*, $f(x)$ has two local extreme values $f(1) = 13$ and $f(2) = 12$.

Meanwhile, the two terminal values are $f(0) = 8$ and $f(3) = 17$.

Therefore, we conclude its global maximal and minimal values in the given interval are $\boxed{17}$ and $\boxed{8}$, respectively.

Done.

For a multi-variable function, partial derivatives against all variables must either equal 0 or not exist at local extremes. This conclusion is obvious because if it does not hold then it is possible to find a new extreme by adjusting one variable and keeping the rest unchanged.

3.18.2 Determine Inflection Points

An inflection point is where the concavity of a curve changes. Therefore, it must occur at places where the second derivative equals 0 or <u>does not exist</u>. If follows that the process of determining inflection points is similar to finding local extremes.

3.18.3 The L'Hôpital Rule

The L'Hôpital rule provides a convenient way to calculate limits in the form of $\frac{0}{0}$. Additionally, this method can also be able to handle limits in the form of $\frac{\pm\infty}{\pm\infty}$, $0 \cdot \pm\infty$, 0^0, and so on, because all of them can be transformed to $\frac{0}{0}$.

Chapter 3: Derivative

> **Theorem 3.18.2 L'Hôpital's Rule**
>
> If $\lim_{x \to c} f(x) = \lim_{x \to c} g(x) = 0$ or $\pm\infty$, $g'(x) \neq 0$ for all x in the neighborhood of c (possibly except $x = c$), and $\lim_{x \to c} \dfrac{f'(x)}{g'(x)}$ exists, then
>
> $$\lim_{x \to c} \frac{f(x)}{g(x)} = \lim_{x \to c} \frac{f'(x)}{g'(x)} \qquad (3.28)$$

Please note that this rule can be applied repeatedly. That is, if $\dfrac{f'(x)}{g'(x)}$ is still one of the qualified forms, *(3.28)* can be applied again which yields

$$\lim_{x \to c} \frac{f(x)}{g(x)} = \lim_{x \to c} \frac{f'(x)}{g'(x)} = \lim_{x \to c} \frac{f''(x)}{g''(x)}$$

As an example, $\lim_{x \to 0} \dfrac{\sin x}{x}$ discussed in *Chapter 2* can be solved by using L'Hôpital rule as follows:

$$\lim_{x \to 0} \frac{\sin x}{x} = \lim_{x \to 0} \frac{\cos x}{1} = 1$$

The L'Hôpital rule can also be used to compute one side limit.

Example 3.18.3

Compute
$$\lim_{x \to 0^+} \sqrt{x} \cdot \ln x$$

Solution

This is in the form of $0 \cdot (-\infty)$. Applying L'Hôpital's rule gives

$$\lim_{x \to 0^+} \sqrt{x} \cdot \ln x = \lim_{x \to 0^+} \frac{\ln x}{x^{-\frac{1}{2}}} = \lim_{x \to 0^+} \frac{x^{-1}}{-\frac{1}{2} x^{-\frac{3}{2}}} = \lim_{x \to 0^+} \left(-2 x^{\frac{1}{2}}\right) = \boxed{0}$$

Done.

Chapter 3: Derivative

It is important to note the requirement that $\lim_{x \to c} \dfrac{f'(x)}{g'(x)}$ must exist. Disregarding this requirement will lead to incorrect results. Let's consider the following example:

$$\lim_{x \to \infty} \frac{x + \sin x}{x}$$

This expression is in the form of $\frac{\infty}{\infty}$. Differentiating both the numerator and denominator gives

$$\lim_{x \to \infty} \frac{1 + \cos x}{1} = 1 + \lim_{x \to \infty} \cos x$$

This limit does not exist. Therefore, the L'Hôpital rule should not be used in order to avoid mistakenly claiming the limit of the original problem does not exist. This is because the limit exists by working on the original expression directly:

$$\lim_{x \to \infty} \frac{x + \sin x}{x} = \lim_{x \to \infty} \left(1 + \frac{\sin x}{x}\right) = 1$$

3.18.4 Root Finding Algorithm

Not all equations have explicit solution formulas. For example, the following equation is solvable but its root cannot be expressed using a closed form formula:

$$e^x = x + 2$$

In practice, such equations are usually solved using specially designed computer algorithms. These algorithms use numerical methods to find approximate solution with very high accuracy such as 10^{-10} or even better. There are many different root finding algorithms available. Most commercial grade ones will use a combination of several basic algorithms to achieve a balance between efficiency and stability.

One basic but highly efficient algorithm is called *Newton method*. This algorithm can be explained using the diagram below:

Chapter 3: Derivative

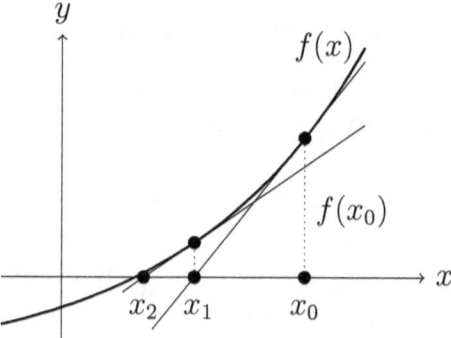

The true root is the x-intercept of $y = f(x)$. Let x_0 be the initial guess. If $|f(x_0)| < \varepsilon$ where ε is a desired accuracy tolerance such as 10^{-10}, then x_0 is declared as the solution, i.e. an approximation to the true root satisfying the desired accuracy. Otherwise, the next candidate x_1 is determined by finding the x-intercept of the tangent line at point (x_0, y_0) where $y_0 = f(x_0)$, i.e.,

$$f'(x_0) = \frac{f(x_0)}{x_0 - x_1} \implies x_1 = x_0 - \frac{f(x_0)}{f'(x_0)} \qquad (3.29)$$

If $|f(x_1)| < \varepsilon$, then x_1 is declared as a solution. Otherwise, the next candidate x_2 is located using the same method. This process can be repeated by using the recursion *(3.30)* below until a solution is found or the number of iterations exceeds a preset limit. The latter case means that this particular solving process has failed.

$$x_{n+1} = x_n - \frac{f(x_n)}{f'(x_n)} \qquad (3.30)$$

The biggest advantage of the Newton method is its efficiency. As shown in the diagram above, the initial value x_0 is quite far from the true root. However, x_2 is already pretty close to it. In fact, the Newton method is one of fastest root finding algorithms. That being said, this method's dependence on first derivative means that if the derivative does not exist at any of the intermediate points, or the derivative equals 0, the Newton method will fail. In these cases, the solving program should switch to an alternative method which may be more tolerant (i.e. more stable) but less efficient.

Example 3.18.4

Solve the following equation numerically:

$$e^x - x^2 = 0$$

The first step to use the Newton method is to find an initial guess x_0. Therefore, it is necessary to determine where a root may locate. A useful and intuitive tool is the intermediate value theorem.

> **Theorem 3.18.3 Intermediate Value Theorem**
>
> If a function $f(x)$ is continuous over a closed interval $[a, b]$, then for any value k between $f(a)$ and $f(b)$, there must exist a point $c \in [a, b]$ such that $f(c) = k$.

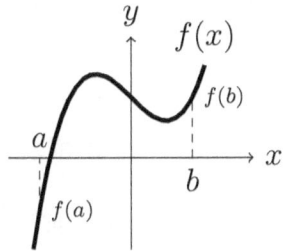

$f(x)$ will take every value between $f(a)$ and $f(b)$ over the interval $[a, b]$ at least once. Therefore, if the signs of $f(a)$ and $f(b)$ are opposite, then there must be a root within $[a, b]$, i.e. $f(c) = 0$, $c \in [a, b]$.

Theorem 3.18.3 has many uses in numerical computing.

Solution to Example 3.18.4

Let $f(x) = e^x - x^2$. Because $f(0) = 1$ and $f(-1) = 1/e - 1 < 0$, therefore this function must have a zero in $[0, 1]$. Any value within this interval can be served as the initial guess x_0. Here, let's use the mid-point $x_0 = -0.5$. Meanwhile, we can set tolerance $\varepsilon = 10^{-6}$. Note that the first derivative $f'(x) = e^x - 2x$.

$x_0 = -0.5$ $\qquad\qquad\qquad\qquad\qquad\quad |f(x_0)| \approx 0.3565 > \varepsilon$
$x_1 = -0.5 - f(-0.5)/f'(-0.5) \approx -0.722 \quad |f(x_1)| \approx -0.03536 > \varepsilon$
$x_2 = \cdots \approx -0.7036 \qquad\qquad\qquad\qquad |f(x_2)| \approx -0.00025 > \varepsilon$
$x_3 = \cdots \approx -0.70347 \qquad\qquad\qquad\quad\; |f(x_3)| \approx 1.3 \times 10^{-8} < \varepsilon$

Chapter 3: Derivative

Hence, we declare a solution as $x = \boxed{-0.70347}$.

Done.

As noted earlier, the Newton method is not always able to find a root. Additionally, because almost all numeric root finding algorithm relies on an initial guess, therefore they may not be able to find all the roots if $f(x) = 0$ has multiple solutions.

3.18.5 Regression and Machine Learning

When the number of independent equations in a system is not equal to the number of independent variables, this system may not have a unique solution. However, it may still be useful to find the most optimal solution according to a given criteria. Regression is such an example.

The target of a linear regression is to find a straight line $y = f(x) = mx + b$ to best fit n points (x_i, y_i) where $i = 1, 2, \cdots, n$. The best fitting should minimize the regression error which is usually defined as:

$$Err = \sum_{i=1}^{n}(y_i - f(x_i))^2 = (y_1 - f(x_1))^2 + \cdots + (y_n - f(x_n))^2$$

Each term, $y_i - f(x_i)$, is the difference between the actual y_i and the prediction produced by this linear model $f(x_i)$. The purpose of squaring every term is to avoid undesirable cancellation between positive and negative differences.

In a linear regression, Err is a function of m and b because all the x_i and y_i are known constants. Therefore, the goal becomes to determine the pair of parameters (m, b) such that the value of the Err function is minimized. In this case, there are two variables, one function and many data points.

Minimizing predication error is the same as finding the most optimal result. Thus, such problems are often referred as optimization problems.

Similar to the root finding algorithms, there are many well-known optimization algorithms as well. One of them is called the Levenberg-Marquardt method which is specialized to minimize a sum of squares such as the *Err* function above. At its core, this method utilizes partial derivatives, such as the Jocobian matrix which is discussed in *Section 3.17*, to help determine the searching direction of the optimal parameters.

As an oversimplified summary, machine learning is a gigantic optimization algorithm. Using the linear regression as a simple example, the collection of (x_i, y_i) is called the training set in machine learning's terminology. The straight line $y = mx + b$ is the working model. Upon having determined the optimal parameters' set (m, b), then computer will be able to predict the value of y when given new input of x. This is an extremely simplified supervised machine learning example. From mathematical point of view, the ability to efficiently and reliably determine the optimal parameter is a key component which almost certainly will involve derivatives.

Chapter 3: Derivative

3.19 Practice

Practice 1

Find the derivative of a^x where a is a constant.

Practice 2

Find the derivative of $\log_a x$ where a is a constant.

Practice 3

Find the derivative of function $\arctan x$.

Practice 4

Compute the derivative of $x \ln x$.

Practice 5

Convert $x^2 + y^2 = R^2$, where R is a constant, to an explicit form and then compute $\frac{dy}{dx}$. Compare this solution with that presented in *Example 3.7.1* on *page 34*.

Practice 6

The equation $x^y = y^x$ describes a curve in the first quadrant of the plane containing the point $P = (4, 2)$. Compute the slope of the line that is tangent to this curve at P.

(Bennett)

Practice 7

Consider the parabola $y = ax^2 + 2019x + 2019$. There exists exactly one circle which is centered on the x-axis and is tangent to the parabola at exactly two points. It turns out that one of these tangent points is $(0, 2019)$. Find a.

(SMT)

Practice 8

Let $f(x)$ be an odd function which is differentiable over $(-\infty, +\infty)$. Show that $f'(x)$ is even.

(UConn)

Practice 9

Let $f_0(x) = (\sqrt{e})^x$, and recursively define $f_{n+1}(x) = f'_n(x)$ for integers $n \geq 0$. Compute $\sum_{k=0}^{\infty} f_k(1)$.

(SMT)

Practice 10

Let curve $\mathbb{C}$ is defined as
$$\begin{cases} x = \cot t \\ y = \dfrac{\cos(2t)}{\sin t} \end{cases}$$
where $t \in (0, \pi)$. Find all inflection points of this curve.

(China)

Chapter 3: Derivative

Practice 11

Compute $\lim\limits_{x \to 0} x \ln x$ and $\lim\limits_{x \to 0} x^x$.

Practice 12

Compute
$$\lim_{x \to 0} \frac{(1 - \cos x)^2}{x^2 - x^2 \cos^2 x}$$

(SMT)

Practice 13

For a given $x > 0$, let a_n be the sequence defined by $a_1 = x$ for $n = 1$ and $a_n = x^{a_{n-1}}$ for $n \geq 2$. Find the largest x for which $\lim\limits_{n \to \infty} a_n$ exists.

(SMT)

Practice 14

Compute the value of
$$\lim_{x \to \pi} \frac{\ln(2 + \cos x)}{(3^{\sin x} - 1)^2}$$

(China)

Practice 15

Let $f(x) = x^2 \cos(ax)$ where a is a constant. Find the 50th order derivative of $f(x)$, i.e. $f^{(50)}(x)$.

(China)

Practice 16

If water is poured into a right cone whose height is H and base's radius is R at a speed of A, what is the speed the water is rising when the depth of water is half of the cone's height?

Practice 17

Use the derivative definition to prove the product rule.

Chapter 3: Derivative

Chapter 4

Integral

4.1 Rectangular Approximation Model

We have already shown in *Section 2.6.3* on *page 19* that the area under a curve can be approximated by using a series of equal-width rectangles. This is called the rectangular approximation model, or RAM as the abbreviation.

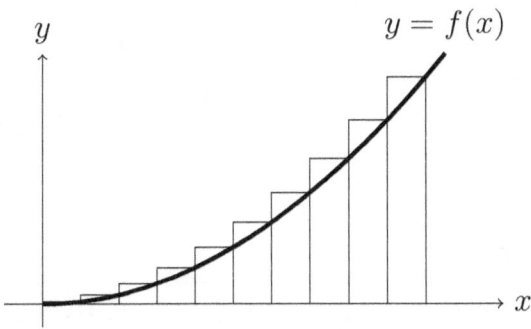

In *Section 2.6.3*, the height of a rectangle is determined by the x coordinate of the right end of its base. This setup is called RAM-R. It can be shown that as long as the function $f(x)$ is continuous, the final results will be the same if the height is determined by the left end (RAM-L), mid-point (RAM-M), or any point on the rectangle's base.

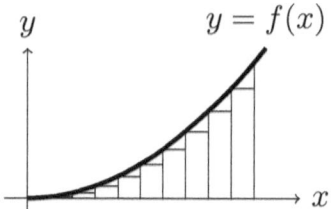 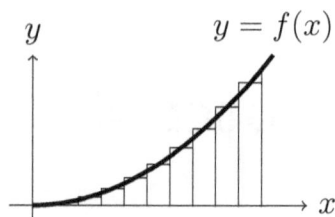

Intuitively, this is because when n approaches infinity, the rectangle's width will approach 0. If $f(x)$ is continuous, then the heights of these rectangles under different setups will approach to the same value.

4.2 Riemann Sum and Integral

Additionally, it turns out that, so long as $f(x)$ is continuous, these rectangles do not need to have equal widths. The area computed using any RAM model will be still the same. To be more precise, let's consider an arbitrary partition $\mathbb{P}$ of a closed $[a, b]$ into n back-to-back sub-intervals, as shown below.

In each sub-interval $[x_{k-1}, x_k]$, a number c_k is chosen and its corresponding functional value $f(c_k)$ is calculated. Then the sum

$$S_n = \sum_{k=1}^{n} f(c_k) \Delta x_k \qquad (4.1)$$

is called the Riemann sum of function $f(x)$ over $[a, b]$. It can be shown that the limit

$$\lim_{\|\mathbb{P}\|\to 0} \sum_{k=1}^{n} f(c_k)\Delta x_k \qquad (4.2)$$

is independent of the construction of $\mathbb{P}$ and choices of c_k. Here, $\|\mathbb{P}\|$, referred as the *norm* or *mesh* of $\mathbb{P}$, denotes the longest length among its sub-intervals.

The result of *(4.2)* is called *Riemann integral* and is written as

$$\lim_{\|\mathbb{P}\|\to 0} \sum_{k=1}^{n} f(c_k)\Delta x_k = \int_a^b f(x)\mathrm{d}x \qquad (4.3)$$

The right side of the above relation means to sum the products $f(x)\,\mathrm{d}x$ as the value of x progresses from a to b. Because $\mathrm{d}x \to 0$, so will be each product $f(x)\,\mathrm{d}x$, therefore the integral is to sum an infinite number of "zero" terms. While the sum of a finite number of zeros must be zero, the sum of an infinite number of zeros can be any value. In this case, this sum will equal the area under a curve.

Similar to the notation that $\mathrm{d}x$ can be viewed as Δx as it approaches 0, the integral $\int$ can be viewed as the sum of an infinite number of terms $\sum_{k=1}^{\infty}$ when each term becomes infinitely small.

Riemann integral is not the only way by which integral can be defined, but it is the most widely used. Unless explicitly specified otherwise, the term integral usually means Riemann integral.

It is important to point out that because the value of $f(c_k)$ (or $f(x)$ on the right side) in *(4.1)* can be negative, so will be the product $f(c_k)\Delta x_k$ (or $f(x)\,\mathrm{d}x$). This is corresponding to the case when a curve lays beneath the x-axis. In such cases, the rectangles constructed in RAM will be formed downward and their areas will be treated as negative. Accordingly, an integral can be negative. For example,

$$\int_0^{\pi} \sin x\,\mathrm{d}x = -\int_{\pi}^{2\pi} \sin x\,\mathrm{d}x \implies \int_0^{2\pi} \sin x\,\mathrm{d}x = 0$$

Chapter 4: Integral

because the area under this curve in $[\pi, 2\pi]$ is negative and cancels the positive area in $[0, \pi]$.

4.3 Fundamental Theorem of Calculus

The value of an integral can be computed using the RAM model. However, it is possible to obtain the result directly by "inversing" the derivative operator, i.e. introducing an anti-derivative function.

Assuming there exists a differentiable function $F(x)$ satisfying

$$\mathrm{d}F(x) = f(x)\mathrm{d}x \qquad (4.4)$$

then $F(x)$ is called an anti-derivative of $f(x)$. Because the derivative of a constant is zero, therefore if $F(x)$ satisfies *(4.4)*, so will be all the functions in the form of $F(x) + C$ where C is an arbitrary constant. This means that anti-derivative function is not unique. Instead, it is a family of functions which differ from each other by just some constants.

With the help of anti-derivative functions, an integral can be computed directly. This is stated below:

> **Theorem 4.3.1   Integral Evaluation Theorem**
>
> If $f(x)$ is continuous over $[a, b]$ and $F(x)$ is <u>any</u> anti-derivative of $f(x)$, then
>
> $$\int_a^b f(x)\,\mathrm{d}x = F(b) - F(a) \qquad (4.5)$$

Because all anti-derivatives differ only by constants, therefore the difference on the right of *(4.5)* will always be the same regardless of which particular anti-derivative is used.

Equation 4.5 is often written as

$$\int_a^b f(x)\,dx = F(x)|_a^b = F(b) - F(a)$$

Accordingly, computing an integral requires two steps

i) Finding an anti-derivative $F(x)$ of the given function $f(x)$

ii) Evaluate the difference by setting the lower and upper terminals using (4.5)

If an integral's terminals are not given, then it is called indefinite integral which is the same as finding the anti-derivatives. Hence, the result of an indefinite integral is a family of functions which differ only by some constants C, i.e.

$$\int f(x) = F(x) + C$$

where C is any constant. Sometimes, the statement "where C is any constant" may be omitted to keep the solution concise when there is no confusion. Despite it is a commonly used practice, it is strongly recommended to always include this statement in a formal exam for rigorousness reason.

An integral with terminal values can be explicitly called as a definite integral. The result of a definite integral is a numerical value.

Let's consider an example.

Example 4.3.1

Find the area under the curve $y = x^2$ between $x = 0$ and 1.

Solution

By *(2.8)* on *page 19*, the area will be

$$\lim_{n\to\infty} S_n = \lim_{n\to\infty} \sum_{k=1}^{n} \frac{1}{n}\left(\frac{k}{n}\right)^2$$

Chapter 4: Integral

By the definition of (4.1), S_n above is a Riemann sum where function $f(x) = x^2$. Then, by the definition of Riemann integral (4.3), its value can be computed using

$$\lim_{n \to \infty} \sum_{k=1}^{n} \frac{1}{n} \left(\frac{k}{n}\right)^2 = \int_0^1 x^2 \, dx$$

Now, because $\left(\frac{1}{3}x^3\right)' = x^2$, we find $F(x) = \frac{1}{3}x^3$ is one antiderivative of $f(x)$. Hence, by (4.5), we have

$$\int_0^1 x^2 \, dx = \left(\frac{1}{3}x\right)\Big|_0^1 = \left(\frac{1}{3} \cdot 1^3\right) - \left(\frac{1}{3} \cdot 0^3\right) = \boxed{\frac{1}{3}}$$

This result agrees with the conclusion obtained earlier in (2.8).

Done.

Equation 4.5 is a very important relation and is also called the second form of the fundamental theorem of calculus. The first form of the fundamental theorem of calculus is given below.

Theorem 4.3.2   Fundamental theorem of Calculus

If function $f(x)$ is continuous over $[a, b]$, then the function

$$F(x) = \int_a^x f(t) \, dt$$

is differentiable everywhere in $[a, b]$ and

$$\frac{dF}{dx} = \frac{d}{dx} \int_a^x f(t) \, dt = f(x) \qquad (4.6)$$

Combining (4.5) and (4.6) clearly shows that derivative and integral are inverse to each other: if the integral of $f(x)$ is $F(x)$, then the derivative of $F(x)$ is $f(x)$.

The following relation can be intuitively understood by using the inverse relation because a pair of the $\int$ and d operators can be canceled.

$$\int dx = x + C \qquad (4.7)$$

Please note that this interpretation is just to conveniently explain the relationship between integral and derivative. A more rigid way is shown below:

$$\int dx = \int 1\, dx \quad \text{and} \quad (x)' = 1 \implies \int dx = x + C$$

The proof of the fundamental theorem of calculus *(4.6)* can be illustrated geometrically.

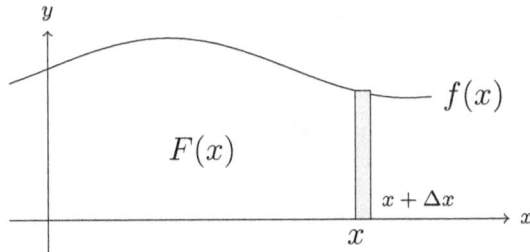

By definition, $F(x)$ represents that area beneath the curve $f(x)$ over the interval $[a, x]$. Now, let's consider a slim rectangle of width Δx at the its ending point x. There are two ways to compute its area S. On one hand, it is the difference between an enlarged area beneath $f(x)$ and the original one, i.e.

$$S = F(x + \Delta x) - F(x)$$

On the other hand, it can be approximated as the product of its height and base:

$$S \approx f(x)\Delta x$$

Relating these two relations gives

$$F(x + \Delta x) - F(x) \approx f(x)\Delta x \implies f(x) \approx \frac{F(x + \Delta x) - F(x)}{\Delta x}$$

Chapter 4: Integral

Approximation becomes equality when $\Delta x \to 0$, i.e.

$$f(x) = \lim_{\Delta x \to 0} \frac{F(x + \Delta x) - F(x)}{\Delta x} = \frac{\mathrm{d}F}{\mathrm{d}x}$$

The last step uses the definition of derivative *(3.1)* on *page 25*.

4.4 The Substitution Method

Finding anti-derivatives is the basic task in integral calculation. However, it is usually more challenging than determining derivatives. The basic formulas are those listed in the table of derivatives which can be found in *Section 3.14* on *page 44*. The challenge is how to decompose a given function into something readily integrateable using the table of derivatives.

The most basic technique is the substitution method. It appears in the vast majority of integration calculations. Below are a couple of examples.

Example 4.4.1

Compute $\int \cos 2x \, \mathrm{d}x$.

Solution

Let $u = 2x$, then $\mathrm{d}u = 2 \, \mathrm{d}x$. It follows that

$$I = \int \cos u \left(\frac{1}{2} \mathrm{d}u\right) = \frac{1}{2} \int \cos u \, \mathrm{d}u = \frac{1}{2} \sin u + C = \frac{1}{2} \sin 2x + C$$

It can also be written implicitly:

$$\int \cos 2x \, \mathrm{d}x = \int \cos 2x \left(\frac{1}{2} \mathrm{d}2x\right) = \frac{1}{2} \int \cos 2x \, \mathrm{d}2x = \frac{1}{2} \sin 2x + C$$

Done.

As shown, this substitution method can be viewed as the counterpart of the chain rule in derivative calculation. The goal is to transform the part inside the d operator the same as the variable in front. For example, in the previous example, because the to-be-integrated function is $\cos 2x$, it is necessary to convert $\mathrm{d}x$ to $\mathrm{d}(2x)$.

It is suggested that student should get use to the implicit form when using this method, i.e. the 2nd style solution in the previous example.

Example 4.4.2

Compute
$$\int \frac{1}{a^2 + x^2}\,\mathrm{d}x$$

Solution

The given expression looks similar to the derivative of $\arctan x$:

$$\mathrm{d}(\arctan x) = \frac{1}{1+x^2}\,\mathrm{d}x$$

Hence, we have

$$\int \frac{1}{a^2+x^2}\,\mathrm{d}x = \int \frac{1}{a^2\left(1+\left(\frac{x}{a}\right)^2\right)} \cdot a\left(\mathrm{d}\left(\frac{x}{a}\right)\right)$$

$$= \frac{1}{a}\int \frac{1}{1+\left(\frac{x}{a}\right)^2}\,\mathrm{d}\left(\frac{x}{a}\right)$$

$$= \boxed{\frac{1}{a}\arctan\left(\frac{x}{a}\right) + C} \qquad (4.8)$$

Done.

Equation 4.8 can be used as a well-known result.

4.5 Trigonometric Substitution

As a special substitution, trigonometric substitution plays an important role and widely used in integral computation. Unlike the regular substitution method which is discussed in the previous section, the trigonometric substitution is to convert the given function to a trigonometric expression for easier manipulation by using various trigonometric identities.

Let's first consider an example.

Example 4.5.1

Compute
$$I = \int_0^1 x\sqrt{1-x^2}\,dx$$

Solution

Because $x \in [0, 1]$, therefore it is possible to substitute x with $\sin t$ where $t \in [0, \frac{\pi}{2}]$. Then

$$\sqrt{1-x^2} = \cos t \quad \text{and} \quad dx = d\sin t = \cos t\,dt$$

It follows that

$$\begin{aligned}
I &= \int_0^{\frac{\pi}{2}} \sin t \cos t \cos t\,dt \\
&= \int_0^{\frac{\pi}{2}} \sin t \cos^2 t\,dt \\
&= \int_0^{\frac{\pi}{2}} \sin t(1 - \sin^2 t)\,dt \\
&= \int_0^{\frac{\pi}{2}} (\sin t - \sin^3 t)\,dt
\end{aligned}$$

By the triple angle formula, we have

$$\sin 3t = 3\sin t - 4\sin^3 t \implies \sin^3 t = \frac{1}{4} \times (3\sin t - \sin 3t)$$

Setting this to the above integral gives

$$\begin{aligned}
I &= \int_0^{\frac{\pi}{2}} \left(\sin t - \frac{1}{4} \times (3\sin t - \sin 3t)\right) dt \\
&= \frac{1}{4} \int_0^{\frac{\pi}{2}} (\sin t + \sin 3t)\, dt \\
&= \frac{1}{4} \left(\int_0^{\frac{\pi}{2}} \sin t\, dt + \int_0^{\frac{\pi}{2}} \sin 3t\, dt \right) \\
&= \frac{1}{4} \left(\int_0^{\frac{\pi}{2}} \sin t\, dt + \frac{1}{3} \int_0^{\frac{\pi}{2}} \sin 3t\, d3t \right) \\
&= \frac{1}{4} \left(-\cos t - \frac{1}{3} \cos 3t \right) \Big|_0^{\frac{\pi}{2}} \\
&= \boxed{\frac{1}{3}}
\end{aligned}$$

Done.

As shown in the previous example, it is imperative to be familiar with various trigonometric identities in order to use this method effectively. The book *Trigonometry* in Math All Star series has an in-depth discussion of trigonometric identities.

There are several commonly used substitutions:

$$\begin{aligned}
|x| \leq 1 &\implies x = \sin t \text{ or } x = \cos t \\
x \in \mathbb{R} &\implies x = \tan t \text{ or } x = \cot t \\
|x| \geq 1 &\implies x = \sec t \text{ or } x = \csc t
\end{aligned}$$

After the initial substitution, the next step is usually to use the definitions of these trigonometric functions to simplify the expression. Then, various trigonometric identities can be used to transform the expression to a form that can be readily integrated. As a general rule, it is usually easy to integrate a one-degree expression than to tackle a higher degree one. This is already shown in *Example 4.5.1* where the term $\sin^3 t$ is transformed to a difference of two one-degree terms using the triple angle formula. As such, the double and triple angle formulas listed below will be useful.

$$\sin^2 x = \frac{1}{2}(1 - \cos(2x)), \qquad \cos^2 x = \frac{1}{2}(1 + \cos(2x))$$
$$\sin^3 x = \frac{1}{4}(3\sin x - \sin(3x)), \qquad \cos^3 x = \frac{1}{4}(3\cos x + \cos(3x))$$

4.6 Integration By Parts

Integration by parts is the inverse to the derivative's product rule. Let both u and v are two functions with respect to variable x. Then the product rule states that

$$\frac{d}{dx}(u(x)v(x)) = u(x)\frac{d}{dx}v(x) + v(x)\frac{d}{dx}u(x)$$

Integrating both sides gives

$$u(x)v(x) = \int u(x)\frac{d}{dx}v(x) + \int v(x)\frac{d}{dx}u(x)$$

Re-arranging the above relation yields

Theorem 4.6.1   Integration by Parts

$$\int u(x)\,dv(x) = u(x)v(x) - \int v(x)\,du(x) \qquad (4.9)$$

Equation 4.9 can be written in a more concise way as

$$\int uv' = uv - \int vu' \qquad (4.10)$$

In essence, this method permits to first break a product into two parts, and then to swap out the hard-to-be-integrated one. This method is often being considered when handling the following two forms:

- $\int xf(x)\,dx$ where the anti-derivative of $f(x)$ is obtainable

- $\displaystyle\int e^x f(x)\,\mathrm{d}x$

This is because, in the first case, x will diminish after swap-out. To see this, let the anti-derivative of $f(x)$ be $F(x)$. Then

$$\int xf(x)\,\mathrm{d}x = \int x\,\mathrm{d}F(x) = xF(x) - \int F(x)\,\mathrm{d}x$$

In the second case, e^x can be directly transferred into the d part because $(e^x)' = e^x$:

$$\int e^x f(x)\,\mathrm{d}x = \int f(x)\,\mathrm{d}e^x = e^x f(x) - \int e^x\,\mathrm{d}f(x)$$

Both cases will present in the examples below.

Example 4.6.1

Evaluate
$$\int x\cos x\,\mathrm{d}x$$

Solution

Let $u = x$ and $v = \sin x$. Then $v' = \cos x\,\mathrm{d}x$. Hence,

$$\begin{aligned}
\int x\cos x\,\mathrm{d}x &= \int x\,\mathrm{d}\sin x \\
&= x\sin x - \int \sin x\,\mathrm{d}x \\
&= \boxed{x\sin x + \cos x + C}
\end{aligned}$$

This result can be verified by noting

$$(x\sin x + \cos x)' = (x\cos x + \sin x) - \sin x = x\cos x$$

Done.

Sometime, it is necessary to repeatedly apply this technique to find the final result. Here is an example.

Chapter 4: Integral

Example 4.6.2

Evaluate
$$\int x^2 e^x \, dx$$

Solution

Applying integration by part the first time:

$$\int x^2 e^x \, dx = \int x^2 \, d(e^x) = x^2 e^x - \int e^x \, d(x^2) = x^2 e^x - 2 \int x e^x \, dx$$

The remaining integral can be tackled using the same technique:

$$\int x e^x \, dx = \int x \, d(e^x) = x e^x - \int e^x \, dx = x e^x - e^x + C$$

Combining both gives the final result as

$$\int x^2 e^x \, dx = \boxed{(x^2 - 2x + 2)e^x + C}$$

Done.

As shown in the previous example, because $(e^x)' = e^x$, therefore the effect of applying the integration by parts to $\int e^x f(x) \, dx$ is to continuously simplify $f(x)$ till the final results can be obtained, i.e.:

$$\int x^2 e^x \, dx \to \int x e^x \, dx \to \int e^x \, dx$$

4.7 Additional Techniques

The substitution and the integrate by part are undoubtedly the must-known basic integrating methods. Additionally, there are many other integrating techniques. Some frequently used ones are given in this section.

4.7.1 Watch Out Absolute Value

When finding anti-derivatives, one detail requiring attention is the function domain. This means that sometimes it may be necessary to taking an absolute value operator. Let's consider the integral of $\int \frac{1}{x}\,dx$. The correct answer is

$$\int \frac{1}{x}\,dx = \ln|x| + C \qquad (4.11)$$

This is because the domain of the logarithm function is positive numbers. Therefore $|x|$ must be used instead of x in order to ensure this. It can be shown that the result will still be correct even if the interval given when calculating a definite integral includes negative values.

Example 4.7.1

Compute
$$\int_{-2}^{-1} \frac{1}{x}\,dx$$

Solution

By *(4.11)*, we have

$$\int_{-2}^{-1} \frac{1}{x}\,dx = (\ln|x|)\big|_{-2}^{-1} = -\ln 2$$

This result can be verified by letting $y = -x$, then we have $y \in [1, 2]$ which is always positive. It follows that

$$\int_{-2}^{-1} \frac{1}{x}\,dx = \int_{2}^{1} \frac{1}{-y}\,d(-y) = \ln y\big|_{2}^{1} = -\ln 2$$

Done.

The alternative solution in this example, i.e. the y-substitution, also shows that it is not necessary for the upper terminal to be

Chapter 4: Integral

greater than the lower one. When these two terminals are switched, the resulting value will be negated. In another word, there is no requirement that $b \geq a$ in the integral evaluation theorem (4.5) on page 70.

4.7.2 The Symmetry Method

The technique illustrated in the next example is a symmetry based substitution. Its goal is not to find the answer directly. Instead, it aims to find a mirroring expression with equal value. Then, the value of the original integral must equal to the half of the sum of itself and its mirroring expression. This idea is similar to the symmetry and bijection methods in combinatorics which are discussed in the books *Counting* and *More on Counting*.

Example 4.7.2

Evaluate
$$\int_0^\pi \frac{x \sin x}{1 + \cos^2 x} \, dx$$

(Utah)

Solution

Let $z = \pi - x$, then $dz = -dx$. It follows that

$$\int_0^\pi \frac{x \sin x}{1 + \cos^2 x} dx$$
$$= -\int_\pi^0 \frac{(\pi - z)\sin(\pi - z)}{1 + \cos^2(\pi - z)} dz$$
$$= \int_0^\pi \frac{\pi \sin z}{1 + \cos^2 z} dz - \int_0^\pi \frac{z \sin z}{1 + \cos^2 z} dz$$

The last part is the same as the to-be-evaluated expression.

$$\therefore \quad \int_0^\pi \frac{x \sin x}{1 + \cos^2 x} dx = \frac{1}{2} \int_0^\pi \frac{\pi \sin x}{1 + \cos^2 x} dx \qquad (4.12)$$

The right side of (4.12) can be evaluated as:

$$\frac{\pi}{2}\int_0^\pi \frac{\sin x}{1+\cos^2 x}dx = \frac{\pi}{2}\int_0^\pi \frac{d(-\cos x)}{1+\cos^2 x}$$

$$= \frac{\pi}{2}\left(\arctan(\cos x)|_\pi^0\right)$$

$$= \frac{\pi}{2}\left(\frac{\pi}{4}-\left(-\frac{\pi}{4}\right)\right)$$

$$= \boxed{\frac{\pi^2}{4}}$$

Done.

4.7.3 The Symmetry Method (II)

Another form of the symmetry method is to solve a pair of integrals together. These two integrals are usually symmetric expressions and convertible to each other via integration by parts.

Example 4.7.3

Evaluate

$$\int e^{ax}\cos(bx)\,dx \quad \text{and} \quad \int e^{ax}\sin(bx)\,dx$$

Solution

Applying integration by parts on the first expression gives

$$\int e^{ax}\cos(bx)\,dx = \frac{1}{a}\int \cos(bx)\,d(e^{ax})$$

$$= \frac{1}{a}e^{ax}\cos(bx) - \frac{1}{a}\int e^{ax}\,d(cos(bx))$$

$$= \frac{1}{a}e^{ax}\cos(bx) + \frac{b}{a}\int e^{ax}\sin(bx)\,dx \quad (4.13)$$

Chapter 4: Integral

Similarly, we can have

$$\int e^{ax} \sin(bx) \, dx = \frac{1}{a} e^{ax} \sin(bx) - \frac{b}{a} \int e^{ax} \cos(bx) \, dx \qquad (4.14)$$

Now, solving the two equations (4.13) and (4.14) will find the final results as

$$\begin{cases} \int e^{ax} \cos(bx) \, dx = \boxed{\dfrac{e^{ax}}{a^2 + b^2} \left(b \sin(bx) + a \cos(ax)\right) + C} \\ \int e^{ax} \sin(bx) \, dx = \boxed{\dfrac{e^{ax}}{a^2 + b^2} \left(a \sin(bx) - b \cos(ax)\right) + C} \end{cases}$$

Done.

It is possible to continue (4.13) to get the final answer without explicitly solving the second integral as an independent problem. However, it still involves the second integral implicitly.

4.7.4 Special Pattern

The problem below is from the Stanford Math Tournament. It can be solved using the substitution method. However, a neater solution exists by exploiting the fact that the derivative of a square root is proportional to its reciprocal, i.e.:

$$\sqrt{x} = \frac{1}{2} \cdot \frac{1}{\sqrt{x}}$$

The to-be-evaluated integral exhibits both features of square root and reciprocal. Therefore, it may be possible to find its antiderivative directly by try-and-error.

Example 4.7.4

Compute the value of

$$\int_0^2 \left(\sqrt{\frac{4-x}{x}} - \sqrt{\frac{x}{4-x}}\right) dx$$

Solution

It can be verified that

$$\left(\sqrt{x}\sqrt{4-x}\right)' = \frac{1}{2}\left(\sqrt{\frac{4-x}{x}} - \sqrt{\frac{x}{4-x}}\right)$$

Therefore, the given integral must equal

$$\left(2\sqrt{x}\sqrt{4-x}\right)\Big|_0^2 = \boxed{4}$$

Done.

4.7.5 Partial Fraction Decomposition

Partial fraction decomposition refer to the technique of breaking a fraction into the sum of several simpler fractions. For example,

$$\frac{1}{x^2 - a^2} = \frac{1}{(x+a)(x-a)} = \frac{1}{2a}\left(\frac{1}{x-a} - \frac{1}{x+a}\right)$$

This technique is useful not only in solving pre-calculus competition problems, but also in calculus. Let's consider an example.

Example 4.7.5

Compute

$$\int \frac{1}{x^2 - a^x} dx$$

Chapter 4: Integral

Solution

The form $x^2 - a^2$ may be tackled using trigonometric substitution. However, in this case, using partial fraction decomposition is more appropriate.

$$\int \frac{1}{x^2 - a^2}\,dx = \frac{1}{2a} \int \left(\frac{1}{x-a} - \frac{1}{x+a} \right) dx$$

$$= \frac{1}{2a} \left(\int \frac{1}{x-a}\,dx - \int \frac{1}{x+a}\,dx \right)$$

$$= \frac{1}{2a} \left(\ln|x-a| - \ln|x+a| \right) + C$$

$$= \boxed{\frac{1}{2a} \ln\left| \frac{x-a}{x+a} \right| + C}$$

The integrating of the two terms in the third step uses substitutions $dx = d(x-a)$ and $dx = d(x+a)$, respectively.

Done.

If the to-be-integrated function is a quotient of two polynomials and its denominator can be factorized, then this partial fraction decomposition method can always be used to at least simplify the problem. In such cases, the following results can be useful:

$$\int \frac{1}{x \pm a}\,dx = \ln|x \pm a| + C \tag{4.15}$$

$$\int \frac{1}{x^2 - a^2}\,dx = \frac{1}{2a} \ln\left| \frac{x-a}{x+a} \right| + C \tag{4.16}$$

$$\int \frac{1}{x^2 + a^2}\,dx = \frac{1}{a} \arctan \frac{x}{a} + C \tag{4.17}$$

The last result is obtained in *Example 4.4.2 on page 75.*

4.7.6 Completing the Square

When the denominator of the quotient is in the form of x^2+bx+c where $b, c \neq 0$ and cannot be factorized, it may be possible to use the completing the square method. Let's consider an example.

Example 4.7.6

Compute
$$\int \frac{x+1}{x^2+x+1} dx$$

Solution

Let's first transform the denominator:
$$\int \frac{x+1}{x^2+x+1} dx = \int \frac{x+1}{\left(x+\frac{1}{2}\right)^2 + \frac{3}{4}} dx$$

Then, letting $t = x + \frac{1}{2}$ yields:

$$\int \frac{t + \frac{1}{2}}{t^2 + \frac{3}{4}} dt = \int \frac{t}{t^2 + \frac{3}{4}} dt + \frac{1}{2} \int \frac{1}{t^2 + \frac{3}{4}} dt$$

$$= \frac{1}{2} \int \frac{d\left(t^2 + \frac{3}{4}\right)}{t^2 + \frac{3}{4}} + \frac{\sqrt{3}}{3} \int \frac{d\left(\frac{2}{\sqrt{3}}t\right)}{1 + \left(\frac{2}{\sqrt{3}}t\right)^2}$$

$$= \frac{1}{2} \ln\left(t^2 + \frac{3}{4}\right) + \frac{\sqrt{3}}{3} \arctan\left(\frac{2}{\sqrt{3}}t\right) + C$$

$$\boxed{= \frac{1}{2} \ln\left(x^2 + x + 1\right) + \frac{\sqrt{3}}{3} \arctan \frac{2x+1}{\sqrt{3}} + C}$$

Done.

The result of this type of problems is usually a combination of a logarithm function and an inverse-tangent function which are

Chapter 4: Integral

corresponding to *(4.15)* and *(4.17)* given in the previous section, respectively.

4.7.7 The Construction Method

Some problems require the construction method which uses various transformation to construct an easy-to-integrate form. Most of these transformations are not obvious and, thus, require experience.

Example 4.7.7

Compute
$$\int \frac{1}{\sin x}\,\mathrm{d}x$$

Solution

Multiplying both the numerator and the denominator by $\sin x$:
$$\int \frac{\sin x}{\sin^2 x}\,\mathrm{d}x = \int \frac{-\mathrm{d}\cos x}{1 - \cos^2 x} = \int \frac{\mathrm{d}\cos x}{\cos^2 x - 1}$$

Then by *(4.16)* or partial fraction decomposition, we have
$$\int \frac{\mathrm{d}\cos x}{\cos^2 x - 1} = \frac{1}{2}\ln\left|\frac{1-\cos x}{1+\cos x}\right| + C$$

Now by the double angle formulas listed at the end of *Section 4.5* on *page 76*
$$1 - \cos x = 2\sin^2 \frac{x}{2} \quad \text{and} \quad 1 + \cos x = 2\cos^2 \frac{x}{2}$$

the previous result can be further simplified as
$$\frac{1}{2}\ln\left|\frac{1-\cos x}{1+\cos x}\right| + C = \frac{1}{2}\ln \tan^2 \frac{x}{2} + C = \boxed{\ln\left|\tan \frac{x}{2}\right| + C}$$

Done.

This problem can also be solved using other ways of construction. Here are two alternative solutions.

Alternative Solution

Let $x = 2u$, then $dx = 2\,du$. Also note the identities that $\sin 2u = 2 \sin u \cos u$ and $1 = \sin^2 u + \cos^2 u$. Therefore the original integral is equivalent to

$$\int \frac{1}{\sin x}\,dx = \int \frac{1}{\sin(2u)}\,d(2u)$$
$$= \int \frac{2}{2 \sin u \cos u}\,du$$
$$= \int \frac{\sin^2 u + \cos^2 u}{\sin u \cos u}\,du$$
$$= \int \left(\frac{\sin u}{\cos u} + \frac{\cos x}{\sin u}\right) du$$
$$= -\int \frac{d\cos x}{\cos x} + \int \frac{d\sin x}{\sin u}$$
$$= -\ln|\cos x| + \ln|\sin x| + C$$
$$= \ln|\tan u| + C$$
$$= \boxed{\ln\left|\tan \frac{x}{2}\right| + C}$$

Done.

Alternative Solution 2

This transformation is even less obvious.
$$\int \frac{1}{\sin x}\,dx = \int \csc x\,dx = \int \csc x \cdot \frac{\csc x + \cot x}{\csc x + \cot x}\,dx$$

Now, note the following two results by the chain rule and the quotient rule: (They can be regarded as basic derivative formulas for those advanced students.)

$$\frac{d}{dx} \csc x = \frac{d}{dx}\left(\frac{1}{\sin x}\right) = \frac{-\cos x}{\sin^2 x} = -\csc x \cot x$$

Chapter 4: Integral

$$\frac{d}{dx}\cot x = \frac{d}{dx}\left(-\frac{\cos x}{\sin x}\right) = \frac{\sin x \sin x - \cos x \cos x}{\sin^2 x} = -\csc^2 x$$

It follows that the derivative of the original integral's denominator is equal to

$$\frac{d}{dx}(\csc x + \cot x) = -\csc x \cot x - \csc^2 x$$

which equals to the opposite of its numerator. Hence, the original integral equals

$$-\int \frac{d(\csc x + \cot x)}{\csc x + \cot x} = \boxed{-\ln|\csc x + \cot x| + C}$$

It can be shown that this result is equivalent to the one obtained in the previous two solutions by trigonometric transformation.

Done.

Therefore, we have

$$\boxed{\int \csc x = \ln\left|\tan\frac{x}{2}\right| + C = \ln|\csc x + \cot x| + C} \qquad (4.18)$$

Similarly, it can be shown that

$$\boxed{\int \sec x = \ln|\sec x + \tan x| + C} \qquad (4.19)$$

Both *(4.18)* and *(4.19)* may be used as well-known results when solving complex integral problems.

4.7.8 Recursion

When the to-be-integrated function involves n^{th} power, it may be possible to employ integration by parts to obtain a recursion.

This method is usually seen in proofs, but sometimes may appear in calculation problems too.

Example 4.7.8

Given a constant $a \neq 0$, evaluation

$$J_n = \int \frac{\mathrm{d}x}{(x^2 + a^2)^n}$$

Solution

Applying integration by parts gives:

$$J_n = \int \frac{\mathrm{d}x}{(x^2 + a^2)^n}$$

$$= \frac{x}{(x^2 + a^2)^n} - \int x \mathrm{d}\left(\frac{1}{(x^2 + a^2)^n}\right)$$

$$= \frac{x}{(x^2 + a^2)^n} + 2n \int \frac{x^2}{(x^2 + a^2)^{n+1}} \mathrm{d}x$$

Now because

$$\int \frac{x^2}{(x^2 + a^2)^{n+1}} \mathrm{d}x = \int \frac{x^2 + a^2 - a^2}{(x^2 + a^2)^{n+1}} \mathrm{d}x$$

$$= \int \frac{\mathrm{d}x}{(x^2 + a^2)^n} - a^2 \int \frac{\mathrm{d}x}{(x^2 + a^2)^{n+1}}$$

$$= J_n - a^2 J_{n+1}$$

Therefore, we have

$$J_n = \frac{x}{(x^2 + a^2)^n} + 2n \left(J_n - a^2 J_{n+1}\right)$$

or

$$J_{n+1} = \frac{1}{2na^2} \left((2n - 1) J_n + \frac{x}{(x^2 + a^2)^n}\right) \qquad (4.20)$$

When $n = 1$, we have

$$J_1 = \int \frac{\mathrm{d}x}{x^2 + a^2} = \frac{1}{a} \arctan \frac{x}{a} + C$$

Therefore, it is possible to use the recursion *(4.20)* to find J_n. In this case, J_n's general form is quite complex and thus will be omitted because the purpose of this example is to illustrate the use of recursion in integration.

Done.

4.8 Improper Integral

There are several types of improper integrals. Most of them are intuitive to understand and easy to handle. The first type is when one or both terminals are infinities:

$$\int_a^\infty f(x) \, \mathrm{d}x \qquad (4.21)$$

Recall that integral can be treated as a special form of summation. Therefore, *(4.21)* is similar in nature as the problem appeared in *Example 2.2.1* on *page 6*:

$$S = \sum_{k=1}^\infty \frac{1}{k(k+1)} \qquad (4.22)$$

This expression can be computed by first defining

$$S_n = \sum_{k=1}^n \frac{1}{k(k+1)}$$

and then compute the limit as n approaches infinity:

$$S = \lim_{n \to \infty} S_n = \lim_{n \to \infty} \sum_{k=1}^n \frac{1}{k(k+1)}$$

In a similar way, *(4.21)* can be computed as

$$\int_a^\infty f(x)\,\mathrm{d}x = \lim_{b\to\infty} \int_a^b f(x)\,\mathrm{d}x \qquad (4.23)$$

Next is an example of this kind.

Example 4.8.1

Compute
$$\int_0^\infty e^{-x}\,\mathrm{d}x$$

Solution

First, let's compute

$$\int_0^b e^{-x}\,\mathrm{d}x = -\int_0^b e^{-x}\,\mathrm{d}(-x) = -e^{-x}\Big|_0^b = 1 - \frac{1}{e^b}$$

Then applying *(4.23)* yields

$$\int_0^\infty e^{-x}\,\mathrm{d}x = \lim_{b\to\infty}\left(1 - \frac{1}{e^b}\right) = \boxed{1}$$

Quite often, the solution can be short-handed written as below if there is no confusion:

$$\int_0^\infty e^{-x}\,\mathrm{d}x = -e^{-x}\Big|_0^\infty = 1$$

Done.

Another example of improper integral is when there exists an infinite discontinuity in the interval. In this case, the original integral can be broke into two parts at the point of discontinuity.

Chapter 4: Integral

Example 4.8.2

Compute
$$\int_0^3 \frac{1}{(x-1)^{\frac{2}{3}}}\,dx$$

Solution

The given function is discontinuous at $x = 1$, but continuous over both $[0, 1)$ and $(1, 3]$. Therefore, we can first integrate over these two sub-intervals separately and then add up their results.

$$\begin{aligned}
\int_0^1 \frac{1}{(x-1)^{\frac{2}{3}}}\,dx &= \lim_{b \to 1^-} \int_0^b \frac{1}{(x-1)^{\frac{2}{3}}}\,dx \\
&= \lim_{b \to 1^-} \int_0^b (x-1)^{-\frac{2}{3}}\,d(x-1) \\
&= \lim_{b \to 1^-} 3(x-1)^{\frac{1}{3}}\Big|_0^b \\
&= \lim_{b \to 1^-} \left(3(b-1)^{\frac{1}{3}} - (-3)\right) \\
&= 3
\end{aligned}$$

Similarly,
$$\int_1^3 \frac{1}{(x-1)^{\frac{2}{3}}}\,dx = 3\sqrt[3]{2}$$

Hence, the final result is $\boxed{3 + 3\sqrt[3]{2}}$.

Done.

Please note that, this his particular case, the result will be the same if the point of discontinuity is ignored, i.e.

$$\int_0^3 \frac{1}{(x-1)^{\frac{2}{3}}}\,dx = 3\,(x-1)^{\frac{1}{3}}\Big|_0^3 = 3\sqrt[3]{2} + 3$$

However, this is just an coincidence because the left and right limits at the point of discontinuity happen to be the same in this problem. Additionally, using the fundamental theorem of calculus requires the function to be continuous over the to-be-integrated interval.

4.9 Differential Equation

An equation involving a function and its derivatives is called a differential equation. One example is

$$f'(x) = kf(x) \tag{4.24}$$

where k is a constant. The solution to a differential equation is a function. For example, the solution to (4.24) is

$$f(x) = \lambda e^{kx}$$

where λ is any constant. This solution can be directly verified:

$$f'(x) = \left(\lambda e^{kx}\right)' = k\lambda e^{kx} = kf(x)$$

Differential equation plays a vital role in calculus' real life applications such as in physics. A complete literature of solving various types of differential equations is beyond the scope of this book. Here, let's explain how (4.24) is solved as an introductory example and then discuss a couple of other most seen elementary differential equations.

4.9.1 Separable Differential Equation

Let $y = f(x)$. Then (4.24) can be written as

$$\frac{dy}{dx} = ky$$

Chapter 4: Integral

The next step is to separate all the variables. This means to put y and x at different sides of this equation:

$$\frac{dy}{y} = k\,dx$$

Then it is possible to integrate both sides[1]:

$$\int \frac{dy}{y} = k \int dx \implies \ln y + C_1 = kx + C_2 \implies y = e^{kx+(C_2-C_1)}$$

Now, letting $\lambda = e^{C_2-C_1}$ gives the final solution:

$$y = e^{C_2-C_1} e^{kx} = \boxed{\lambda e^{kx}}$$

Because it is possible to separate x and y in (4.24), therefore such equations are called *separable differential equations*.

4.9.2 Integrating Factor

Another commonly seen elementary differential equation solving technique involves *integrating factor*. A classical example which can be solved using this technique is equations in the following form:

$$y'(x) + p(x)y(x) = q(x) \tag{4.25}$$

where $y(x)$ is the to-be-solved function, $p(x)$ and $q(x)$ two known functions of x. To solve (4.25), let

$$u(x) = e^{\int p(x)dx} \tag{4.26}$$

which is the integrating factor for equation (4.25). It can be verified that $u(x)$ satisfies the following relation:

$$u'(x) = p(x)u(x) \tag{4.27}$$

---

[1] Recall that it is possible to differentiate both sides of an equation. It is also possible to integrate both sides of an equation.

This is a separable differential equation which is discussed in the previous section, and *(4.26)* is its solution.

Now, multiplying both sides of *(4.25)* by $u(x)$ will find the left side of the equation is in the form of derivatives' product rule. In fact, transforming the left side into such a form is exactly the purpose of introducing the integrating factor.

$$u(x)y'(x) + \underbrace{u(x)p(x)}_{u'(x)} y(x) = u(x)q(x)$$

$$\therefore \quad (u(x)y(x))' = u(x)q(x) \tag{4.28}$$

Integrating both sides of *(4.28)* and rearranging the result yield

$$y(x) = \frac{\int u(x)q(x)\,\mathrm{d}x + C}{u(x)} \tag{4.29}$$

where C is any constant and $u(x)$ is defined by *(4.26)*.

Equations in different forms may require different integrating factors. However, the target is the same which is to make both sides directly integrable.

4.9.3 Homogeneous Equation

An differential equation in the following form where $a_0, a_1, \cdots, a_n$ are all constants is called homogeneous equation.

$$a_0 y + a_1 y' + a_2 y'' + \cdots + a_n y^{(n)} = 0 \tag{4.30}$$

To solve *(4.30)*, let $y = e^{kx}$ where k is a constant:

$$a_0 e^{kx} + a_1 k e^{kx} + a_2 k^2 e^{kx} + \cdots + a_n k^n e^{kx} = 0$$

Because e^{kx} never equals 0, therefore *(4.30)* is equivalent to

$$a_0 + a_1 k + a_2 k^2 + \cdots + a_n k^n = 0$$

Chapter 4: Integral

which means k must be a root of the following equation:

$$a_0 + a_1 z + a_2 z^2 + \cdots + a_n z^n = 0$$

Then, all the solutions to (4.30) can be written as a combination of all these roots depending on whether or not these roots are distinct. Let's consider the case of second-order as an example.

In order to solve the equation

$$y'' + my' + ny = 0 \qquad (4.31)$$

the first step is to find zeros to

$$z^2 + mz + n = 0$$

Let them be z_1 and z_2.

i) If $z_1 \neq z_2$, then all the solutions to (4.31) are

$$y = c_1 e^{z_1 x} + c_2 e^{z_2 x} \qquad (4.32)$$

where c_1 and c_2 are two arbitrary constants. Please note that z_1 and z_2 can be either real numbers or complex numbers.

ii) If $z_1 = z_2$, then all the solutions are

$$y = (c_1 + c_2 x) e^{z_1 x} \qquad (4.33)$$

where c_1 and c_2 are two arbitrary constants.

Both (4.32) and (4.33) can be validated by setting them back to (4.31). Note that, in the seconding case, it must hold that $m^2 - 4n = 0$ and $z_1 = z_2 = -\frac{m}{2}$.

Solving this type of differential equations shares much similarities with linear recursion using characteristic equations. The latter is discussed in the book *Competitive Algebra*.

4.10 Examples and Applications

The essence of integral is the ability to sum an infinite number of infinitely small quantities. Therefore, many applications of integral first model a problem using infinitely small pieces and then sum them up using integral in order to get the final result.

4.10.1 Compute Arc Length

In order to compute the length of an arc, it is necessary to write the length of a tiny segment of the arc dl using a proper calculus expression. Then, the total length is just the integral of dl.

The expression of dl varies depending on the system used. Under the Descartes coordinates, it is

$$dl = \sqrt{(dx)^2 + (dy)^2} = \sqrt{1 + \left(\frac{dy}{dx}\right)^2}\, dx \qquad (4.34)$$

This result is obtained by approximating a tiny arc using its corresponding chord whose length can be calculated by the Pythagorean theorem. When the chord's length approaches 0, its length will approach that of the arc.

The expression of dl using the polar coordinates is more straightforward and simpler. This is shown below where the notion $r(\theta)$ emphasis that r may be function of θ:

$$dl = r(\theta)\, d\theta \qquad (4.35)$$

If parametric function is used, then dl can be written as *(4.36)* below. This result is derived from *(4.34)* above.

Chapter 4: Integral

$$dl = \sqrt{\left(\frac{dx}{dt}\right)^2 + \left(\frac{dy}{dt}\right)^2} \, dt \qquad (4.36)$$

Because a curve can be expressed in multiple ways under different systems. Therefore, multiple approaches may exist for the same problem. Their results should always be the same. But complexities in deriving these results may vary significantly. Hence, it is important to choose the most appropriate solution.

Let's review an example.

Example 4.10.1

Derive the circumference formula of a circle using two different approaches.

Solution

Let the equation for the circle be $x^2 + y^2 = R^2$. Applying the implicit derivative as shown in *Example 3.7.1 on page 34* gives

$$\frac{dy}{dx} = -\frac{x}{y}$$

Then the length of the arc in the first quadrant equals

$$\frac{C}{4} = \int_0^R \sqrt{1 + \left(\frac{dy}{dx}\right)^2} \, dx$$

$$= \int_0^R \sqrt{1 + \left(-\frac{x}{y}\right)^2} \, dx$$

$$= \int_0^R \sqrt{\frac{y^2 + x^2}{y^2}} \, dx$$

$$= \int_0^R \frac{R}{y} \, dx$$

$$= R \int_0^R \frac{1}{\sqrt{R^2 - x^2}} \, dx$$

Let $x = R\sin\theta$, then $\mathrm{d}x = R\cos\theta$, hence

$$\frac{C}{4} = R\int_0^{\frac{\pi}{2}} \frac{R\cos\theta}{R\cos\theta}\,\mathrm{d}\theta = R\int_0^{\frac{\pi}{2}} \mathrm{d}\theta = \frac{\pi}{2}R \implies C = \boxed{2\pi R}$$

The same result can also be derived using *(4.35)*. Under the polar coordinates, this circle's equation is $r = R$. Hence, its circumference equals

$$\int_0^{2\pi} R\,\mathrm{d}\theta = \boxed{2\pi R}$$

Done.

4.10.2 Compute Area Using Polar Coordinates

As discussed early, Riemann integral originates from area computation. In addition to Descartes system, integral can also be used to calculate areas which is described using polar coordinates. In this case, the area of a slice of section can be expressed as

$$\mathrm{d}s = \frac{1}{2}r^2(\theta)\mathrm{d}\theta \qquad (4.37)$$

Note that r may be a function of θ. *Equation 4.37* can be derived by thinking the slice of section is part of a circle when $\mathrm{d}\theta$ is small. In this case, its area is proportional to that of the whole circle whose radius is r:

$$\mathrm{d}S = \pi r^2 \cdot \frac{\mathrm{d}\theta}{2\pi} = \frac{1}{2}r^2\,\mathrm{d}\theta$$

Some curves, such as a cardioid which is shown in *Example 4.10.2* below, are best described using polar coordinates. Hence, using *(4.37)* is the most convenient way to calculate its area.

Chapter 4: Integral

Example 4.10.2

Compute the area enclosed by the cardioid $r = (1 + \cos\theta)$.

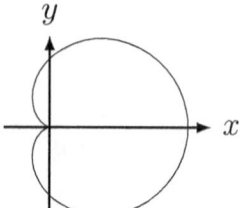

Solution

By *(4.37)*, we have

$$S = \int_0^{2\pi} \frac{1}{2} \times (1 + \cos\theta)^2 \, d\theta$$
$$= \frac{1}{2} \int_0^{2\pi} (1 + 2\cos\theta + \cos^2\theta) \, d\theta$$
$$= \frac{1}{2} \int_0^{2\pi} \left(1 + 2\cos\theta + \frac{1 + \cos 2\theta}{2}\right) d\theta$$
$$= \frac{1}{4} \int_0^{2\pi} (3 + 4\cos\theta + \cos 2\theta) \, d\theta$$
$$= \frac{1}{4} \left(3\theta + 4\sin\theta + \frac{1}{2}\sin 2\theta\right)\Big|_0^{2\pi}$$
$$= \boxed{\frac{3}{2}\pi}$$

Done.

4.10.3 Compute Volume

Integral is also a powerful tool to calculate the volume and surface area of a solid object. The idea underpinning volume calculation is similar to that of area computation. Imaging the diagram below is a vertical cross-section of the target object (e.g. a sphere).

Chapter 4: Integral

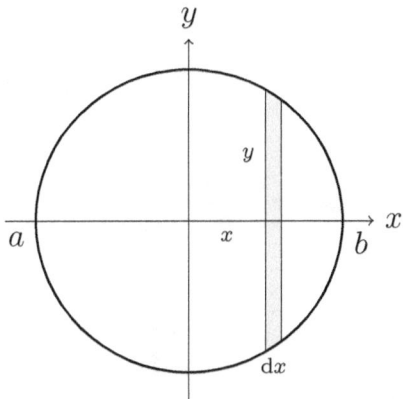

Figure 4.1: volume calculation

Then, the shaded area corresponds to a thin slice of the object which is perpendicular to the x-axis. Let the area of its base be $S(x)$, then the volume of this thin slice will be

$$\mathrm{d}V = S(x)\,\mathrm{d}x$$

where $S(x)$ emphasizes that S may be a function of x. It follows that the volume of this object V equals

$$V = \int_a^b S(x)\,\mathrm{d}x \tag{4.38}$$

where a and b are the boundaries of x. As a special case, if the object is rotational symmetric with respect to x, then the the base of the slice will be a circle whose area $S(x) = \pi y^2$.

Example 4.10.3

Derive the formula of sphere volume.

Solution

Let the radius be R, then the circle is

$$x^2 + y^2 = R^2 \implies y = \sqrt{R^2 - x^2}$$

Therefore, $S = \pi y^2 = \pi(R^2 - x^2)$ and

$$V = \int_{-R}^{R} \pi(R^2 - x^2)\,\mathrm{d}x = \pi\left(R^2 x - \frac{1}{3}x^3\right)\Big|_{-R}^{R} = \boxed{\frac{4\pi}{3}R^3}$$

Done.

Most volume calculation problems involve objects which are rotational symmetric to one of the axes. In such cases, it is always possible to choose a circle as the base which means $S(x)$ is easy to compute. In other cases when $S(x)$ cannot be easily computed, double integral may be necessary for volume computation. The inner integral is used to compute $S(x)$ and then the outer integral is to calculate the volume.

Double integral can be tackled in a similar way as double summation. There is one such problem in the practice.

$$\iint f(x,y)\,\mathrm{d}y\,\mathrm{d}x = \int \left(\int f(x,y)\,\mathrm{d}y\right)\mathrm{d}x \Leftrightarrow \sum_x \sum_y f(x,y)$$

4.10.4 Compute Surface Area

Surface area can be computed in a similar way as volume does. In this case, it is necessary to first compute the side surface area of a thin slice. This area equals the product of its base's perimeter and the arc's length[2]. The arc's length can be computed using either (4.34) or (4.35).

If the object is rotational symmetric with respect to the x-axis, then the base of the slice is a circle. Accordingly, the base's perimeter equals $2\pi y$.

---

[2] Note the difference between computing the volume and the surface area. The former needs to use the height of the slice, i.e. $\mathrm{d}x$. But the latter needs to use the "width" of the ring, i.e. $\mathrm{d}l$.

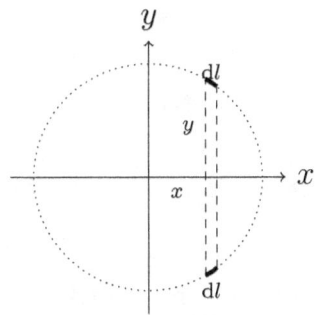

If the object is rotationally symmetric to the x-axis:

$$dS = (2\pi y)\, dl \qquad (4.39)$$

Let's conciser an example.

Example 4.10.4

Derive the formula of calculating a sphere's surface area whose radius is R.

Solution

Using Descartes coordinates, we have

$$A = \int_{-R}^{R} \underbrace{(2\pi y)\sqrt{1 + \left(\frac{dy}{dx}\right)^2}\, dx}_{dl}$$

Because $x^2 + y^2 = R^2$ and $\frac{dy}{dx} = -\frac{x}{y}$ (see *Example 3.7.1* on *page 34*), therefore

$$A = 2\pi \int_{-R}^{R} y\sqrt{1 + \left(-\frac{x}{y}\right)^2}\, dx = 2\pi \int_{-R}^{R} R\, dx = \boxed{4\pi R^2}$$

Under the polar system, $y = R\sin\theta$ and the arc length equals $R\, d\theta$. It follows that

$$A = \int_0^\pi (2\pi R\sin\theta) R\, d\theta = 2\pi R^2 \left(-\cos\theta\right)\big|_0^\pi = \boxed{4\pi R^2}$$

Done.

Chapter 4: Integral

4.10.5 Determine Center of Mass

Integral is also a generalized tool to locate the center of mass, or centroid, of an arbitrary shaped object. To derive the solution, let's start by introducing the concept of *moment* using discrete objects.

Consider two masses, m_1 and m_2, located on the x-axis whose coordinates are x_1 and x_2, respectively. Let the coordinate of their centroid be $\bar{x}$, then it must hold that

$$(x_1 - \bar{x})m_1 = (\bar{x} - x_2)m_2 \implies \bar{x} = \frac{m_1 x_1 + m_2 x_2}{m_1 + m_2} \tag{4.40}$$

The denominator of (4.40) is the total mass. Its numerator is called moment which measures the system's tendency to rotate around the origin. Therefore, we find the location of the centroid is the quotient of the moment dividing the mass. Intuitively, if there are n masses on the x-axis, then the coordinate of their mass is

$$\bar{x} = \frac{\sum_{k=1}^{n} m_k x_k}{\sum_{k=1}^{n} m_k} \tag{4.41}$$

Now, let's consider an area which is bounded by $f(x)$ and $g(x)$, ($a \leq x \leq b$) as shown in the diagram below. Meanwhile, let's also assume its density function is $\rho(x, y)$.

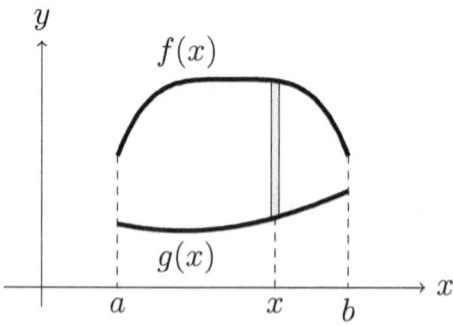

In order to compute its mass and x-moment, let's consider a thin slice which is perpendicular to the x-axis. Its mass will be

$$\mathrm{d}M = \left(\int_{g(x)}^{f(x)} \rho(x,y)\,\mathrm{d}y \right) \mathrm{d}x \qquad (4.42)$$

This is because the mass of each tiny rectangular piece on this slice equals the product of its density $\rho(x,y)$ and its area $\mathrm{d}x\,\mathrm{d}y$. Then, y will progress from $g(x)$ to $f(x)$. Accordingly, the total mass of this whole area can be computed as x progress from a to b.

$$M = \int_a^b \left(\int_{f(x)}^{g(x)} \rho(x,y)\,\mathrm{d}y \right) \mathrm{d}x \qquad (4.43)$$

Meanwhile, the moments of this vertical slice is

$$\mathrm{d}M_y = x \left(\int_{g(x)}^{f(x)} \rho(x,y)\,\mathrm{d}y \right) \mathrm{d}x$$

This means that the total moments will be

$$M_y = \int_a^b x \left(\int_{f(x)}^{g(x)} \rho(x,y)\,\mathrm{d}y \right) \mathrm{d}x \qquad (4.44)$$

Finally, the x coordinate of the centroid is

$$\bar{x} = \frac{M_y}{M} = \frac{\int_a^b x \left(\int_{f(x)}^{g(x)} \rho(x,y)\,\mathrm{d}y \right) \mathrm{d}x}{\int_a^b \left(\int_{f(x)}^{g(x)} \rho(x,y)\,\mathrm{d}y \right) \mathrm{d}x} \qquad (4.45)$$

In many problems, the mass is usually assumed to be uniformly distributed. In another word, the density is a constant. Let it be $\rho(x,y) = \rho$, then

$$\int_{f(x)}^{g(x)} \rho(x,y)\,\mathrm{d}y = \rho \int_{f(x)}^{g(x)} \mathrm{d}y = \rho\left(g(x) - f(x)\right)$$

Setting this to (4.45) leads to

$$\bar{x} = \frac{M_x}{M} = \frac{\int_a^b x\left(g(x) - f(x)\right)\mathrm{d}x}{\int_a^b \left(g(x) - f(x)\right)\mathrm{d}x} \qquad (4.46)$$

Chapter 4: Integral

This numerator of (4.46) can be viewed as the sum of a weighted average of vertically sliced mass.

In order to compute $\bar{y}$, we just need to calculate its moment M_x because the total mass will stay the same. In this case, the moment of this vertical slice will be

$$dM_x = \left(\int_{f(x)}^{g(x)} y \cdot \rho(x,y) \, dy \right) dx$$

Accordingly the total moments will equal

$$M_x = \int_a^b \left(\int_{f(x)}^{g(x)} y \cdot \rho(x,y) \, dy \right) dx \qquad (4.47)$$

If the density is a constant ρ, then (4.47) becomes

$$M_x = \rho \int_a^b \frac{1}{2} \left(g^2(x) - f^2(x) \right) dx \qquad (4.48)$$

It follows that the y-coordinates of the centroid equals

$$\bar{y} = \frac{\int_a^b \frac{1}{2} \left(g^2(x) - f^2(x) \right) dx}{\int_a^b \left(g(x) - f(x) \right) dx} \qquad (4.49)$$

The numerator of this result can be intuitively understood by rewriting is inside part as

$$\underbrace{\frac{1}{2} \left(g(x) + f(x) \right)}_{A} \cdot \underbrace{\left(g(x) - f(x) \right)}_{B} dx$$

Part B is the mass of this slice. Part A is the the mid-point of this slice which is its centroid if the density of this slice is uniformly distributed.

Equation 4.46 and *(4.49)* are the most used in solving math centroid problems.

Chapter 4: Integral

Example 4.10.5

Locate the center of mass of the shape defined by $y = \sin x$ and $y = 0$ over the interval $[0, \pi]$.

Solution

Without loss of generality, let's assume its density $\rho = 1$. Then the total mass of this area equals

$$M = \int_0^\pi \sin x \, dx = -\cos x \Big|_0^\pi = 2$$

By the principle of symmetry, the x-coordinates of its center of mass must be $\frac{\pi}{2}$. But let's verify it using *(4.46)* and *(4.49)*.

$$M_y = \int_0^\pi x (\sin x - 0) \, dx$$
$$= -\int_0^\pi x \, d\cos x$$
$$= -(x \cos x)\Big|_0^\pi + \int_0^\pi \cos x \, dx \quad \text{(integration by parts)}$$
$$= \pi + \sin x \Big|_0^\pi$$
$$= \pi$$

$$M_x = \frac{1}{2} \int_0^\pi \left(\sin^2 x - 0^2\right) dx$$
$$= \frac{1}{2} \int_0^\pi \frac{1 - \cos 2x}{2} \, dx$$
$$= \frac{1}{4} \left(x - \frac{1}{2} \sin 2x\right)\Big|_0^\pi$$
$$= \frac{\pi}{4}$$

Therefore the coordinates of its center of mass are

$$(x, y) = \left(\frac{M_y}{M}, \frac{M_x}{M}\right) = \boxed{\left(\frac{\pi}{2}, \frac{\pi}{8}\right)}$$

Done.

Chapter 4: Integral

4.10.6 Derivative and Integral in Physics

Calculus is an vital tool in physics. Without calculus, it is only possible to study *average*, not *instantaneous* quantities. For example, average speed v.s. instantaneous speed.

Let the distance traveled, speed, and acceleration be S, v, and a, respectively. Then,

$$S = \bar{v}T = \int_0^T v(t)\,dt \tag{4.50}$$

where $\bar{v}$ indicates the average speed during this period and $v(t)$ is instantaneous speed at the moment t. Conversely, the instantaneous speed at moment t can be calculated using derivative:

$$v(t) = \frac{dS}{dt} \tag{4.51}$$

Similarly, the relation between speed and acceleration can be expressed as following where v_0 is the initial speed:

$$v = v_0 + \int_0^T a(t)dt \tag{4.52}$$

Combining *(4.50)* and *(4.52)* gives

$$S = \int_0^T v(t)\,dt = \int_0^T \left(v_0 + \int_0^t a(u)\,du\right) dt \tag{4.53}$$

If the acceleration is a constant a, then *(4.53)* will yield

$$S = \int_0^T \left(v_0 + a\int_0^u du\right) dt = v_0 T + \frac{1}{2}aT^2 \tag{4.54}$$

This result can also be interpreted geometrically on a speed vs time graph. Because the acceleration is a constant, therefore the speed curve is a straight line. The slope of the speed curve is the acceleration. The area beneath the speed curve represents the distance traveled.

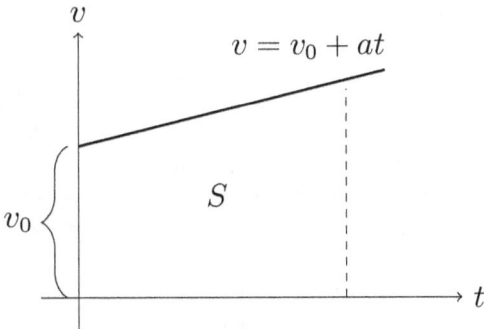

4.10.7 Differential Equation in Physics

Many physics theorems are described using differential equations. For example, the Newton's law of cooling states that the rate of change of an object's temperature is proportional to the difference between temperatures of its own and its surrounding's. In many cases, the surrounding's temperature can be treated as a constant. Then, this law can be described using the following differential equation:

$$f'(t) = k\left(f(t) - C\right) \tag{4.55}$$

where $f(t)$ is the temperature of this object at moment t, k is a constant coefficient, and C is the surrounding temperature.

This equation can be transformed to *(4.24)* on *page 95* by letting

$$g(t) = f(t) - C \implies g'(t) = f'(t) \implies g'(t) = kg(t)$$

It follows that the solution to this differential equation is *(4.56)* below by the conclusion presented in *Section 4.9*:

$$g(t) = \lambda e^{kt} \implies f(t) = \lambda e^{kt} + C \tag{4.56}$$

where λ is a constant.

Now, let's review a 2015 UConn's calculus competition problem:

Chapter 4: Integral

Example 4.10.6

A cup of coffee cools from 164° to 140° (all temperatures are Fahrenheit) in five minutes, and then from 140° to 122° in the next five minutes. What is the room temperature?

Solution

Let the coffer temperature at t minutes from now be $f(t)$ and C be the room temperature. Then by *(4.56)*, we have

$$f(t) = \lambda e^{kt} + C$$

where λ and k are two constants. Setting $t = 0$, 5, and 10, respectively, gives

$$\begin{cases} f(0) &= \lambda + C &= 164 & (a) \\ f(5) &= \lambda e^{5k} + C &= 140 & (b) \\ f(10) &= \lambda e^{10k} + C &= 122 & (c) \end{cases}$$

Equation (a) implies

$$C = 164 - \lambda$$

Setting this to (b) and (c) gives

$$\begin{cases} \lambda e^{5k} - \lambda &= -24 & (d) \\ \lambda e^{10k} - \lambda &= -42 & (e) \end{cases}$$

Dividing (e) by (d) leads to

$$\frac{e^{10k} - 1}{e^{5k} - 1} = \frac{7}{4} \implies e^{5k} + 1 = \frac{7}{4} \implies e^{5k} = \frac{3}{4}$$

Setting this to (d) gives $\lambda = 96$. And finally, setting λ to (a) gives the final answer

$$C = \boxed{68}$$

Done.

4.11 Practice

Practice 1

Compute
$$\int \frac{1}{ax+b}\,dx$$

Practice 2

Let
$$f(x) = \begin{cases} \cos x &, x \in [-\frac{\pi}{2}, 0) \\ e^x &, x \in [0, 1] \end{cases}$$

Compute
$$\int_{-\frac{\pi}{2}}^{1} f(x)\,dx$$

Practice 3

Compute
$$\int_0^4 \frac{dx}{\sqrt{|x-2|}}$$

(SMT)

Practice 4

Evaluate the following integral by using its geometric meaning first. And then compute it again algebraically using the substitution method.
$$\int_0^1 \sqrt{1-x^2}\,dx$$

Chapter 4: Integral

Practice 5

Compute
$$\int_0^{\frac{\pi}{4}} \frac{1}{\sin x + \cos x}\,dx$$

Practice 6

Compute
$$\int \ln x\,dx$$

Practice 7

Compute
$$\int x^3 \ln x\,dx$$

Practice 8

Compute
$$\int \frac{1}{\sqrt{a^2 - x^2}}\,dx$$

Practice 9

Evaluate
$$\int_0^1 x \arcsin x\,dx$$

(China)

Practice 10

Show that
$$\int_0^\infty \frac{x^2}{1+x^4}\,dx = \int_0^\infty \frac{1}{1+x^4}\,dx$$

Practice 11

Evaluate
$$\int_0^\infty \frac{x^2}{1+x^4}\,dx$$

Practice 12

Derive the volume formula of a cone.

Practice 13

Construct one polynomial $f(x)$ with real coefficients and with all of the following properties:

a) it is an even function.

b) $f(2) = f(-2) = 0$.

c) $f(x) > 0$ when $-2 < x < 2$.

d) $f(x)$ archives maximum at $x = \pm 1$.

(UConn)

Chapter 4: Integral

Practice 14

A right circular cone $\mathbb{C}$ has altitude 40 and a circular base of radius 30 inches. A sphere $\mathbb{S}$ is inscribed in $\mathbb{C}$. Find the volume of the region inside $\mathbb{C}$ which is above $\mathbb{S}$.

Practice 15

Compute
$$\int_0^\pi \frac{2x \sin x}{3 + \cos^2 x} \, dx$$

(SMT)

Practice 16

Compute
$$\lim_{x \to 0} \frac{\int_0^x \sin(xt)^2 \, dt}{x^5}$$

(China)

Practice 17

Calculate
$$\lim_{n \to \infty} \frac{1}{n^2} \sum_{k=1}^n \left(k \sin \frac{k\pi}{n} \right)$$

(China)

Practice 18

Evaluate $\lim\limits_{n\to\infty} S_n$ where

$$S_n = 1 - \frac{1}{2} + \frac{1}{3} - \frac{1}{4} + \frac{1}{5} - \cdots + (-1)^{n-1}\frac{1}{n}$$

Practice 19

Compute

$$I = \iiint_S \frac{dxdydz}{(1+x+y+z)^2}$$

where $S = \{x \geq 0, y \geq 0, z \geq 0, x+y+z \leq 1\}$.

(UConn)

Practice 20

Let $f : \mathbb{R} \to \mathbb{R}$ be a periodic continuous function of period $T > 0$, that is $f(x+T) = f(x)$ holds for any $x \in \mathbb{R}$. Show that

$$\lim_{x\to\infty} \frac{1}{x} \int_0^x f(t)dt = \frac{1}{T} \int_0^T f(t)dt$$

Practice 21

Without explicitly evaluating the integral, show that

$$\lim_{n\to\infty} \int_1^2 \ln^n x \, dx = 0 \quad \text{and} \quad \lim_{n\to\infty} \int_2^3 \ln^n x \, dx = \infty$$

Chapter 4: Integral

Practice 22

Determine the differentiable function $f(x)$ such that

$$f(x)\cos x + 2\int_0^x f(t)\sin t\, dt = x+1$$

(China)

Chapter 5

Infinite Series

5.1 Convergence

An infinite series is a sequence containing an unlimited number of terms. In addition to be used in Riemann integral, infinite series has many other applications. One of the most important applications is in Taylor expansion. This chapter dedicates to the topic of infinite series.

An infinite series $\{a_1, a_2, \cdots, a_n, \cdots\}$ can be convergent or divergent. It is said to be convergent if the limit of its partial sum exists when n approaches infinity, i.e.:

$$\lim_{n \to \infty} S_n = \lim_{n \to \infty} (a_1 + a_2 + a_3 + \cdots + a_n) = L$$

In this case, the above relation can be written as

$$a_1 + a_2 + a_3 + \cdots = \sum_{n=1}^{\infty} a_n = L$$

Determining whether or not an infinite series converges is the focus of this section. They are many convergence testing methods. Let's start with an example which can be judged by applying the definition of convergence.

Chapter 5: Infinite Series

Example 5.1.1

Show that the following infinite series converges:

$$\frac{1}{1\times 2},\ \frac{1}{2\times 3},\ \frac{1}{3\times 4},\ \cdots,\ \frac{1}{n\times (n+1)},\ \cdots \quad (5.1)$$

This is the same as *Example 2.2.1* on *page 6*. Therefore, we find the limit of its partial sum is 1 which means this series converges.

Terms in an infinite series can also involve variables. In such a case, whether or not this series converges may depend on the value of this variable. For example, let's consider the following infinite geometric sequence:

$$1,\ x,\ x^2,\ x^3,\ ,\ \cdots,\ x^n,\ \cdots \quad (5.2)$$

The limit of its partial sum

$$\lim_{n\to\infty} S_n = \lim_{n\to\infty}(1+x+x^2+\cdots+x^n) = \lim_{n\to\infty}\frac{1-x^{n+1}}{1-x}$$

exists if and only if $|x| < 1$. Accordingly, $(-1, 1)$ is called this series' *interval of convergence*. Half of this interval's length is called the *radius of convergence* whose value, in this case, equals 1.

Partial sums of both *(5.1.1)* and *(5.2)* can be written in closed forms. Therefore, it is possible to test their convergence by evaluating corresponding limits directly. However, there are many cases where the partial sums cannot be written in closed forms. As a result, this direct testing method will not work in those cases. As such, additional testing methods are needed. They will be discussed in detail later in this section.

It is worth pointing out that excluding the first k terms from an infinite series will not change the fact whether or not the original series converges. This is because the sum of a finite number of terms is also finite.

Many test methods introduced below require "all the terms" of a

Chapter 5: Infinite Series

series meet certain conditions. As a result of the fact stated in the previous paragraph, such requirements can be met if all the terms after the kth term satisfy the said conditions as long as k is a finite number.

5.1.1 Divergence Test

This is an elimination testing method. In order for an infinite series to converge, the value of its nth term must approach 0 as n approaches infinitely large, i.e.

$$\lim_{n\to\infty} a_n = 0 \qquad (5.3)$$

If this does not hold, then the partial sum of the given sequence will continue changing and not settle on a fixed value.

Let's consider a simple example.

Example 5.1.2

Determine whether or not the series $\{a_n = \frac{2n}{3n+1}\}$ converges.

Solution

This series diverges because

$$\lim_{n\to\infty} \frac{2n}{3n+1} = \frac{2}{3} \neq 0$$

Done.

It is imperative to note that *(5.3)* is just a necessary condition for a series to converge, but not a sufficient condition. A counter example is $\{a_n = 1/\ln(n+1)\}$. Proving its divergence is one problem in this chapter's practice.

Chapter 5: Infinite Series

5.1.2 Comparison Test

This method is similar to the Sandwich theorem and the bounded monotonic function testing discussed in *Chapter 2*.

Let $\{a_n\}$ and $\{b_n\}$ be two infinite series whose terms satisfy $0 \leq a_n \leq b_n$. Then the series $\{a_n\}$ is convergent if $\{b_n\}$ converges. This is because the partial sum of series $\{a_n\}$ is monotonically increasing but eventually bounded by the limit of $\{b_n\}$, therefore its limit must exist.

Example 5.1.3

Show that the following sequence is convergent:

$$\frac{1}{1^2}, \frac{1}{2^2}, \frac{1}{3^2}, \ldots, \frac{1}{n^2}, \ldots$$

Proof

By the conclusion of *Example 5.1.1*, we find the series $\{\frac{1}{n \times (n+1)}\}$ converges. Then because

$$0 < \frac{1}{(n+1)^2} < \frac{1}{n \times (n+1)}$$

holds for all n, we can conclude the given series must converge.

$$QED$$

Correspondingly, if the series $\{a_n\}$ is known to be divergent, then so will be the series $\{b_n\}$. The comparison test is widely used to derive new results from existing conclusions. For example, it is well-known that the series $\{\frac{1}{n}\}$ is divergent (which will be proved later), therefore it will be sufficient to show

$$\frac{1}{\ln(n+1)} > \frac{1}{n}$$

in order to prove $\{1/\ln(n+1)\}$ diverges.

5.1.3 Absolute Convergence Test

A series $\{a_n\}$ is said to be *absolutely convergent* if its corresponding series $\{|a_n|\}$ converges. Correspondingly, a series is said to conditionally converge if it converges but does not converge absolutely. Absolute convergence is stronger than conditional convergence. In fact, it can be shown that an absolutely convergent series must conditionally converge. The proof of this conclusion is left as a practice.

That being said, sometimes it may be easier to show a series converges absolutely than to prove it is conditionally convergent.

Example 5.1.4

Show that the series $\sum_{n=1}^{\infty} \dfrac{\sin n}{n^2}$ converges.

Proof

As the series $\{\frac{1}{n^2}\}$ converges according to *Example 5.1.3* and

$$0 < \sum_{n=1}^{\infty} \left|\frac{\sin n}{n^2}\right| \leq \sum_{n=1}^{\infty} \frac{1}{n^2}$$

we find the given series converges absolutely. This means that this series must converge.

QED

Chapter 5: Infinite Series

5.1.4 Ratio Test

This test is suitable when the ratio of $\dfrac{a_{n+1}}{a_n}$ is easy to compute. It states that if $\lim\limits_{n\to\infty} \left|\dfrac{a_{n+1}}{a_n}\right| = L$ exists, and

i) $L < 1$, then $\{a_n\}$ converges absolutely

ii) $L > 1$, then $\{a_n\}$ diverges

iii) $L = 1$, then the test is inconclusive

Intuitively, the existence of L means that when n is sufficiently large, the term a_n will change geometrically at a speed around L. In another word, all the remaining terms $\{a_n, a_{n+1}, \cdots\}$ will behave like a geometric sequence whose common ratio is approximately L. It is well-known that a geometric sequence will converge if and only if the absolute value of its common ratio is less than 1. Therefore, we reckon the original sequence will converge if $L < 1$ and diverge if $L > 1$. Given it is an approximation to a geometric sequence, therefore the case when $L = 1$ needs to be further examined. In fact, the proof of the ratio test is based on the idea of finding an appropriate geometric sequence. This will be left as a practice.

Let's consider an example to show how this test is used.

Example 5.1.5

Show that the following series converges:

$$\frac{1}{1!}, \frac{1}{2!}, \frac{1}{3!}, \cdots, \frac{1}{n!}, \cdots$$

Proof

Because every term in this series is positive and

$$\lim_{n\to\infty} \left(\frac{1}{(n+1)!} \div \frac{1}{n!}\right) = \lim_{n\to\infty} \frac{1}{n+1} = 0 < 1$$

therefore, we conclude this series converges.

$$QED$$

5.1.5 Root Test

The root test is similar to the ratio test. It is suitable when the term of a series is in an exponential form. This test states that if $\lim_{n \to \infty} \sqrt[n]{|a_n|} = L$ exists, and

i) $L < 1$, then $\{a_n\}$ converges absolutely

ii) $L > 1$, then $\{a_n\}$ diverges

iii) $L = 1$, then the test is inconclusive

The root test can be proved in a similar way as the ratio test.

Example 5.1.6

Show that the series $\left\{ a_n = \left(\dfrac{1}{2n+1} \right)^n \right\}$ converges.

Proof

Because

$$\lim_{n \to \infty} \sqrt[n]{\left(\frac{n}{2n+1} \right)^n} = \lim_{n \to \infty} \frac{n}{2n+1} = \frac{1}{2} < 1$$

therefore, the said series converges.

$$QED$$

5.1.6 Limit Test

Let $\{a_n\}$ and $\{b_n\}$ be two series with positive terms. If $\lim\limits_{n \to \infty} \dfrac{a_n}{b_n} = L > 0$, then these two series either both converge or both diverge.

Intuitively, if L exists and is a non-zero constant, then a_n and b_n will be proportional with a finite constant multiple when n is sufficiently large. In another word, the relation between these two series' terms will become linear when n is sufficiently large. As a result, their partial sum of these terms will be linear too. Hence, the limits of their partial sums will be linear too if exist, or both do not exist.

Example 5.1.3 on *page 122* can also be solved using this test because $\dfrac{1}{n(n+1)}$ converges and

$$\lim_{n \to \infty} \left(\frac{1}{n^2} \div \frac{1}{n \times (n+1)} \right) = \lim_{n \to \infty} \frac{n+1}{n} = 1 > 0$$

5.1.7 Integration Test

Let $f(x)$ be continuous, <u>decreasing</u> and <u>positive</u> for $x \geq 1$. Then $\sum\limits_{n=1}^{\infty} f(n)$ converges if and only if $\int_{1}^{\infty} f(x)\,\mathrm{d}x$ converges.

This test can be explained using the rectangular area approximation model by taking left and right ending points, respectively. Rectangles constructed using RAM-L and RAM-R will be the same but their indexes will be off by 1. This means the first rectangle in the RAM-R is the same as the second rectangle in the RAM-L. The second rectangle in the RAM-R is the same as the third rectangle in the RAM-L, and so on. Accordingly, the total areas under these two models will be

$$S_{RAM-L} = a_1 + a_2 + a_3 + \cdots$$
$$S_{RAM-R} = a_2 + a_3 + a_4 + \cdots$$

Then because $f(x)$ is decreasing and positive, we have

$$0 \leq S_{RAM-R} \leq \underbrace{\int_1^\infty f(x)\,dx}_{I} \leq S_{RAM-L}$$

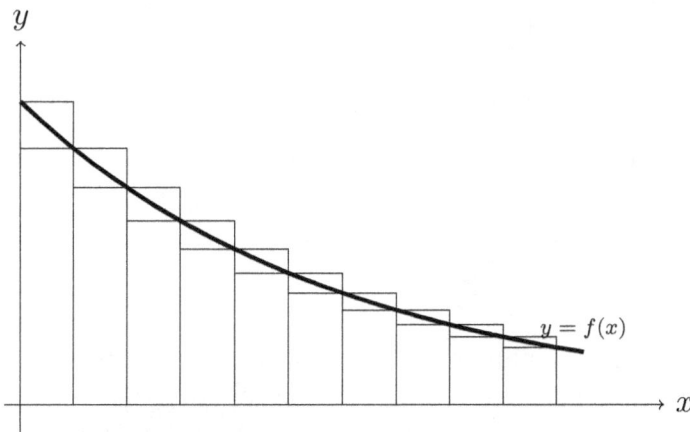

Applying the comparison test concludes that S_{RAM-R} converges if I is convergent and S_{RAM-L} diverges if I is divergent. However, by the principle of the Riemann sum, S_{RAM-L} equals S_{RAM-R} as n approaches infinity, i.e. these two quantities are the same thing. Hence, this integral test holds.

This test will be appropriate if integrating its term is relatively easy. One of its classical applications is to check the convergence of the p-series which is defined below.

Example 5.1.7

Show that the following p-series where p is a positive number converges if and only if $p > 1$:

$$\sum_{n=1}^\infty \frac{1}{n^p} = \frac{1}{1^p} + \frac{1}{2^p} + \frac{1}{3^p} + \frac{1}{4^p} + \cdots$$

Chapter 5: Infinite Series

Proof

Because

$$\int_1^\infty \frac{1}{x^p}\,dx = \begin{cases} \frac{1}{1-p}x^{1-p}\Big|_1^\infty = \frac{1}{p-1} & (p > 1) \\ \ln x\Big|_1^\infty = \infty & (p = 1) \\ \frac{1}{1-p}x^{1-p}\Big|_1^\infty = \infty & (0 < p < 1) \end{cases}$$

thus, by the integration test, the p-series only converges when $p > 1$.

QED

Applying the conclusion of this example, we know that the series $\{\frac{1}{n}\}$ is divergent because $p = 1$, but the series $\{\frac{1}{n^2}\}$ is convergent as $p = 2 > 1$. The p-series is often used as a boundary condition in the comparison test and the limit test.

5.1.8 Alternating Series Test

This test, also referred as *Leibniz Test*, states that if $a_n > 0$ then the alternating series

$$a_1, \ -a_2, \ a_3, \ -a_4, \ a_5, \ -a_6, \ \cdots$$

converges if $a_{n+1} < a_n$ holds for all n and $\lim_{n \to \infty} a_n = 0$.

Intuitively, a series satisfying this condition represents a wavy curve around the x-axis with diminishing amplitudes.

Let's consider an example.

Example 5.1.8

Show that the following series converges

$$1, -\frac{1}{2}, \frac{1}{3}, -\frac{1}{4}, \frac{1}{5}, -\frac{1}{6}, \cdots$$

Proof

Because

$$1, \frac{1}{2}, \frac{1}{3}, \frac{1}{4}, \frac{1}{5}, \frac{1}{6}, \cdots$$

is a decreasing positive series and $\lim\limits_{n\to\infty} \frac{1}{n} = 0$, therefore the given series converges by the alternating series test.

QED

5.2 Taylor Expansion

Among all the applications of infinite series, the Taylor expansion is one of the most important one in calculus and other subjects. Given a continuous function $f(x)$ which is infinitely differentiable at $x = a$, the Taylor expansion states

$$f(x) = f(a) + \frac{f'(a)}{1!}(x-a) + \frac{f''(a)}{2!}(x-a)^2 + \frac{f'''(a)}{3!}(x-a)^3 + \cdots \tag{5.4}$$

In another word, the infinite power series on the right will converge to $f(x)$ within its interval of convergence. From *(5.4)*, it is clear that the $x = a$ is always in the interval of convergence. In fact, it is always the center of this interval.

Chapter 5: Infinite Series

This power series in (5.4) is called the Taylor series. When $a = 0$, the Taylor series becomes the Maclaurin series.

$$f(x) = f(0) + \frac{f'(0)}{1!}x + \frac{f''(0)}{2!}x^2 + \frac{f'''(0)}{3!}x^3 + \cdots \qquad (5.5)$$

Determining the Taylor series of a given function is usually an exercise of differentiating and evaluating $f(x)$.

Example 5.2.1

Find the Taylor expansion of the function $f(x) = \frac{1}{1-x}$ around $x = 0$.

Solution

The answer should be the geometric sequence because for $|x| < 1$, we have

$$\frac{1}{1-x} = 1 + x + x^2 + x^3 + \cdots$$

Let's verify it.

$$f(x) = \frac{1}{1-x} \qquad \Longrightarrow f(0) = 1$$

$$f'(x) = \frac{1}{(1-x)^2} \qquad \Longrightarrow f'(0) = 1$$

$$f''(x) = \frac{2}{(1-x)^3} \qquad \Longrightarrow f''(0) = 2$$

$$f'''(x) = \frac{2 \times 3}{(1-x)^4} \qquad \Longrightarrow f'''(0) = 2 \times 3$$

$$\cdots$$

$$f^{(n)} = \frac{n!}{(1-x)^{n+1}} \qquad \Longrightarrow f^{(n)}(0) = n!$$

Setting these to either *(5.4)* (setting $a = 0$) or *(5.5)* gives

$$\frac{1}{1-x} = 1 + x + x^2 + x^3 + \cdots + x^n + \cdots$$

Done.

Similar to the table of derivatives, the table of Taylor expansion shown below must be memorized too.

$$\frac{1}{1-x} = 1 + x + x^2 + x^3 + x^4 + \cdots \qquad (5.6)$$

$$\frac{1}{1+x} = 1 - x + x^2 - x^3 + x^4 - \cdots \qquad (5.7)$$

$$e^x = 1 + x + \frac{x^2}{2!} + \frac{x^3}{3!} + \frac{x^4}{4!} + \cdots \qquad (5.8)$$

$$\sin x = x - \frac{x^3}{3!} + \frac{x^5}{5!} - \frac{x^7}{7!} + \frac{x^9}{9!} - \cdots \qquad (5.9)$$

$$\cos x = 1 - \frac{x^2}{2!} + \frac{x^4}{4!} - \frac{x^6}{6!} + \frac{x^8}{8!} - \cdots \qquad (5.10)$$

$$\ln(x+1) = x - \frac{x^2}{2} + \frac{x^2}{3} - \frac{x^4}{4} + \cdots \qquad (5.11)$$

All of these expansions are performed at $a = 0$. However, they are valid for any x in their respective interval of convergence. For example, it can be verified that

$$\sum_{n=0}^{\infty} \frac{5^n}{n!} = e^5$$

by setting $x = 5$ in *(5.8)*. (Check it on the Wolframalpha site https://www.wolframalpha.com!)

Chapter 5: Infinite Series

5.3 Deriving Taylor Expansion

The power of the Taylor expansion lays in the claim that all the qualified functions, i.e. continuous and infinitely differentiable at $x = a$, can be represented using a polynomial. This implies that such a function can be studied and manipulated as a polynomial which is usually rather easy. We will give several examples later in this chapter.

The Taylor expansion can also be intuitively explained using the idea of approximating an function using a polynomial. As the approximation gets refined better and better, the Taylor's expansion (5.4) can be eventually obtained.

The starting point is obviously to use the tangent line passing the point $x = a$ which can be represented as a one-degree polynomial $c_0 + c_(x - a)$. This is shown in the left diagram below.

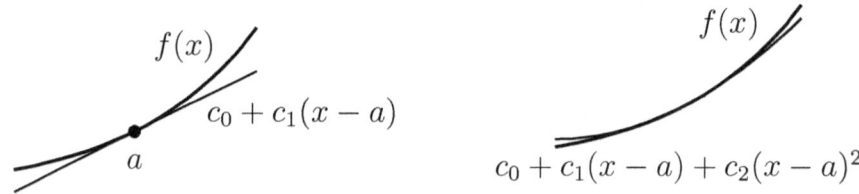

A possible refine is to use a quadratic curve. This is shown as the right diagram above. Generally speaking, the higher the degree the approximating polynomial is, the richer the curve shape can be. Accordingly, the more parameters, i.e. c_i, we have to fit $f(x)$. Let

$$f(x) = c_0 + c_1(x - a) + c_2(x - a)^2 + c_3(x - a)^3 + \cdots$$

Setting $x = a$ gives $c_0 = f(a)$. Next, differentiating both sides:

$$f'(x) = c_1 + 2c_2(x - a)^2 + 3c_3(x - a)^2 + \cdots$$

Setting $x = a$ again gives $c_1 = f'(a)$. Continue this process can solve all the coefficients c_i and lead to (5.4).

5.4 Approximation Error

While mathematics is a precise and rigorous science, the ability to obtain a quantifiable estimation is useful. Separating a quantity into a significant part and an ignorable part allows us to focus on the piece which really matters.

The Taylor's expansion is a great tool to obtain an approximation and a corresponding quantifiable estimation error. This is because the magnitude of each term in the Taylor expansion usually decays quickly. Consequently, it is possible to retain several leading terms as the approximation. The more terms to keep, the more accurate the approximation is. For example, the leading four terms in the polynomial given by (5.9) already resembles the sine function to a high degree around the origin.

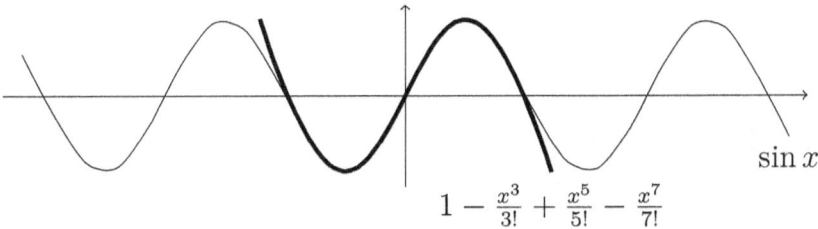

Generally speaking, Taylor expansion can be rewritten as

$$f(x) = f(a) + f'(a)(x-a) + \frac{f''(a)}{2!}(x-a)^2$$
$$+ \frac{f'''(a)}{3!}(x-a)^3 + \cdots + \frac{f^{(n)}(a)}{n!}(x-a)^n + R_n(x) \quad (5.12)$$

where $R_n(x)$, referred as the remainder of order n, satisfies

$$R_n(x) = \frac{f^{(n+1)}(c)}{(n+1)!}(x-a)^{n+1} \quad (5.13)$$

for some c between a and x. Here, $R_n(x)$ is the estimation error for the approximation using the first n terms in the Taylor's expansion.

Note that $R_n(x)$ will be a constant value if x is given. Hence, its maximum is the maximal possible estimation error. Another observation is, as n approaches infinitely large, $R_n(x)$ must approach 0. Obviously, this is necessary for the Taylor expansion to converge to the function $f(x)$.

Another way to write estimation error is the big O notation which is also referred as the Landau's symbol. For example, (5.8) can be written as

$$e^x = 1 + x + \frac{x^2}{2!} + O\left(x^3\right)$$

because the sum of all the remaining terms changes at a speed comparable to x^3. The big O notation is also used in various subjects other than mathematics, such as the computer science. For example, the merger sort algorithm has a complexity of $O(n \log n)$. This means the number of operations used in this algorithm to sort n elements grows at a magnitude of $n \log n$.

5.5 Examples and Applications

5.5.1 Limit Calculation

Representing a function using approximation with a quantifiable error provides another way to calculate the limits of some functions.

Let's revisit *Example 2.3.2* on *page 10*.

Compute the value of $\lim_{n \to \infty} n^2 \left(1 - \cos \frac{\pi}{n}\right)$.

Solution

As n approaches infinitely large, $\frac{\pi}{n}$ will approach zero. Then by

(5.10) on *page 131*, $\cos\frac{\pi}{n}$ can be expanded around 0 as:

$$\cos\frac{\pi}{n} = 1 - \frac{1}{2!}\left(\frac{\pi}{n}\right)^2 + O\left(\frac{1}{n^4}\right)$$

Setting this to the given expression gives:

$$\lim_{n\to\infty} n^2\left(1 - \cos\frac{\pi}{n}\right) = \lim_{n\to\infty}\left(\frac{\pi^2}{2!} + O\left(\frac{1}{n^2}\right)\right) = \boxed{\frac{\pi^2}{2}}$$

because the term $O(\frac{1}{n^2})$ will disappear as n approaches infinity.

Done.

Because the vast majority of limit calculations makes up of functions which can be expanded using Taylor series, therefore this technique can be very useful albeit may be computational heavier than other methods.

5.5.2 Series Differentiation and Integral

Because the Taylor series is convergent, therefore it is acceptable to differentiate and integrate both sides of an expansion. This provides an efficient method to derive new results from known ones.

For example, the identity below

$$\frac{1}{(1-x)^2} = 1 + 2x + 3x^2 + 4x^3 + \cdots \tag{5.14}$$

can be obtained directly by differentiating both sides of *(5.6)*:

$$\frac{1}{1-x} = 1 + x + x^2 + x^3 + \cdots$$

In fact, *(5.11)* in the table of Taylor expansions can be obtained

Chapter 5: Infinite Series

by integrating both sides of (5.7):

$$\frac{1}{1+x} = 1 - x + x^2 - x^3 + \cdots \implies \ln(1+x) = x - \frac{x^2}{2} + \frac{x^3}{3} - \frac{x^4}{4} + \cdots$$

5.5.3 Infinitely Nested Radical

Finding the value of an expression in the following form can be found in various math competitions.

$$\sqrt{1 + \sqrt{1 + \sqrt{1 + \cdots}}} \qquad (5.15)$$

Its solving technique is discussed in the book *Power Calculation*. One approach is to let

$$x = \sqrt{1 + \sqrt{1 + \sqrt{1 + \cdots}}} \qquad (5.16)$$

Then, we have

$$x = \sqrt{1+x} \implies x = \frac{1+\sqrt{5}}{2} \quad (\because x > 0)$$

This solution is acceptable at pre-calculus level especially when only the final result is needed. However, as a matter of rigorousness, two questions must be asked: why the value of (5.15) exists and is unique? Affirmative answers to both questions are necessary prerequisites for (5.16) to hold. This is the starting point of the whole solution.

These two questions are essentially a convergence problem. In order to see this, let sequence $\{a_n\}$ defined recursively as

$$a_{n+1} = \sqrt{1 + a_n}, \qquad a_1 = 1, n \geq 1$$

Then, (5.15) is equivalent to $\lim_{n \to \infty} a_n$. If this limit exists, then it must hold that

$$\lim_{n \to \infty} a_{n+1} = \lim_{n \to \infty} a_n$$

Chapter 5: Infinite Series

Let this limit be x and set it to the recursive definition leads to the equation listed earlier:

$$\lim_{n\to\infty} a_{n+1} = \sqrt{1 + \lim_{n\to\infty} a_n} \implies x = \sqrt{1+x}$$

Therefore, the only missing part in providing a rigorous solution is to show that $\lim_{n\to\infty} a_n$ exists. This can be done by noting that $\{a_n\}$ is strictly increasing and has an upper bound of 2. The existence of the upper bound can be shown using induction: by definition, $a_1 = 1 < 2$. Assuming $a_n < 2$, then

$$a_{n+1} = \sqrt{1+a_n} < \sqrt{1+2} < 2$$

5.5.4 Estimation

Series expansion is a great way to obtain an approximation with reasonable accuracy. For example, when $|x|$ is small, it can be shown that

$$(1+x)^k \approx 1 + kx \tag{5.17}$$

This relations can be intuitively understood by noting the remaining terms in the binomial expansion is $O(x^2)$. When $|x|$ is small, $O(x^2)$ may be ignored. This approximation formula can also be derived using Taylor's expansion. By *(5.4)*, we have

$$f(x) \approx f(a) + f'(a)(x-a)$$

when $(x-a)$ is small. Let $x_0 = a$ and $\Delta x = x - a$. The above relations can be rewritten as

$$f(x_0 + \Delta x) \approx f(x_0) + f'(x_0)\Delta x \tag{5.18}$$

Now let $f(x) = (1+x)^k$, $x_0 = 0$ and $\Delta x = x$, we have

$$f(0+x) \approx f(0) + f'(0)x \implies f(x) \approx 1 + kx$$

which is the same as *(5.17)*. It is worth noting that *(5.18)* is a quite general purposed estimation. In fact, the Newton root finding

Chapter 5: Infinite Series

method *(3.30)* on *page 58* is one of its applications. To see this, let's assume x_0 is an approximation of $f(x)$'s root a. Let $\Delta x = a - x_0$, then by *(5.18)*

$$f(x_0 + \Delta x) \approx f(x_0) + f'(x_0)\Delta x$$

Note that by definition $f(x_0 + \Delta x) = f(a) = 0$, therefore

$$\Delta x \approx -\frac{f'(x_0)}{f(x_0)} \implies a \approx x_0 - \frac{f'(x_0)}{f(x_0)}$$

Letting the left side be x_1 leads to the recursion formula in the Newton's method.

Let's consider another example.

Example 5.5.1

Find an approximate value of $\cos 60.6°$.

Solution

Let $f(x) = \cos x$, $x_0 = 60° = \frac{\pi}{3}$ and $\Delta x = 0.6° = \frac{\pi}{180} \times \frac{6}{10} = \frac{\pi}{300}$. Then by *(5.18)*, we have

$$\cos\left(\frac{\pi}{3} + \frac{\pi}{300}\right) \approx \cos\frac{\pi}{3} + (-\sin\frac{\pi}{3}) \cdot \frac{\pi}{300} = \boxed{0.490931}$$

Done.

A more accurate value is 0.4909037. As this examples shows, the estimation given by a simple polynomial is reasonably accurate.

5.6 Practice

Practice 1

Expand $f(x) = x^4 - 2x^3 + 1$ around $x_0 = 2$.

Practice 2

Estimate the value of $\sqrt[4]{10018}$.

Practice 3

Let $f(x)$ be a twice differentiable continuous function, and $f(0) = f'(0) = 0$, $f''(0) = 6$. Find the value of
$$\lim_{x \to 0} \frac{f(\sin^2 x)}{x^4}$$

(China)

Practice 4

What is the smallest natural number n for which the following limit exists?
$$\lim_{x \to 0} \frac{\sin^n x}{\cos^2 x (1 - \cos x)^3}$$

(SMT)

Chapter 5: Infinite Series

Practice 5

Compute
$$\lim_{x \to 0} \frac{\frac{x^2}{2} + 1 - \sqrt{1+x^2}}{(\cos x - e^{x^2})\sin(x^2)}$$

(China)

Practice 6

Determine if the series $\left\{\frac{n}{e^n}\right\}$ converges.

Practice 7

Show that the inequality $x > \ln(x+1)$ holds for $x > 0$:

Practice 8

Is $\displaystyle\sum_{n=1}^{\infty} \frac{1}{\ln(n+1)}$ convergent?

Practice 9

Compute the limit of the power series below as a rational function in x:

$$1 \cdot 2 + (2 \cdot 3)x + (3 \cdot 4)x^2 + (4 \cdot 5)x^3 + (5 \cdot 6)x^4 + \cdots, \qquad (|x| < 1)$$

(UConn)

Practice 10

Determine the values of α and β such that

$$\lim_{n\to\infty} \frac{n^\alpha}{n^\beta - (n-1)^\beta} = 2020$$

Practice 11

Prove the absolute value test method, i.e. if $\{|a_n|\}$ converges, so will $\{a_n\}$.

Practice 12

Prove the ratio test, i.e. if $\lim\limits_{n\to\infty} \left|\dfrac{a_{n+1}}{a_n}\right| = L$ and $L < 1$, then the series $\{a_n\}$ converges absolutely

Practice 13

Determine whether or not these two series converge:

(A) $\sum\limits_{n=1}^{\infty} \sin\left(\dfrac{\cos n}{n^2}\right)$ (B) $\sum\limits_{n=1}^{\infty} \cos\left(\dfrac{\sin n}{n^2}\right)$

(Bennett)

Chapter 5: Infinite Series

Practice 14

It is well-known that the solution to the Fibonacci sequence is

$$F_n = \frac{1}{\sqrt{5}} \left(\left(\frac{1+\sqrt{5}}{2} \right)^n - \left(\frac{1-\sqrt{5}}{2} \right)^n \right)$$

Show that

$$\lim_{n \to \infty} \frac{F_{n+1}}{F_n} = \frac{1+\sqrt{5}}{2}$$

Practice 15

The oscillation is the time for a pendulum to complete one full back-and-forth movement. Its period T can be computed using the following formula where L is the length of the pendulum and $g \approx 9.8 m/s^2$ is the acceleration of gravity:

$$T = 2\pi \sqrt{\frac{l}{g}}$$

If a particular pendulum's period is 1s, what will be the discrepancy per day if its length is reduced by 1cm due to temperature change?

Appendices

Appendix A

Solutions

Chapter A: Solutions

A.1 Chapter 1

This section is intentionally left blank.

So section numbers of solutions and practices can match.

… Chapter A: Solutions

A.2 Chapter 2

Practice 1

Can the ε-δ definition of limit on page 4 be modified as: for any arbitrary positive number ε, there always exists a number δ such that there are infinitely many $x \in [c-\delta,\ c+\delta]$ satisfying $|f(x) - L| < \varepsilon$?

No. The requirement "always" is different from "infinitely many". A counter example is the sine function. There exist infinitely many x such as $\sin(x) = 0$. However, this function is periodical which does not converge to any value including 0, i.e. $\lim_{x \to \infty} \sin x \neq 0$.

Practice 2

Evaluate
$$\lim_{n \to \infty} \frac{2n^3 + 99n^2 + 1}{3n^3 + n^2 + 3}$$

Dividing both the numerator and denominator by n^3 leads to

$$\lim_{n \to \infty} \frac{2n^3 + 99n^2 + 1}{3n^3 + n^2 + 3} = \lim_{n \to \infty} \frac{2 + \frac{99}{n} + \frac{1}{n^2}}{3 + \frac{1}{n} + \frac{3}{n^2}} = \boxed{\frac{2}{3}}$$

This practice also shows that for a quotient of two polynomials, its limit as n approaches infinity only depends on the terms of highest powers. In this case, they are $2n^3$ and $3n^3$, i.e.

$$\lim_{n \to \infty} \frac{2n^3 + 99n^2 + 1}{3n^3 + n^2 + 3} = \lim_{n \to \infty} \frac{2n^3}{3n^3} = \frac{2}{3}$$

Note that the above relation is not as rigorous as the equation at the beginning of this solution. However, it exhibits an important observation: when n approaches infinity, the term with highest power dominates the growth of the function value. As a result, other terms can be ignored.

Chapter A: Solutions

Practice 3

Use the Sandwich theorem to evaluate

$$\lim_{x \to \infty} \frac{\sin x}{x}$$

Because $-1 \leq \sin x \leq 1$, we always have

$$-\left|\frac{1}{x}\right| \leq \frac{\sin x}{x} \leq \left|\frac{1}{x}\right|$$

Meanwhile, noting that $\lim_{x \to \infty} \left|\frac{1}{x}\right| = 0$ and applying the Sandwich theorem lead to

$$\lim_{x \to \infty} \frac{\sin x}{x} = \boxed{0}$$

Practice 4

Compute the value of

$$\lim_{n \to \infty} (\sqrt{n+1} - \sqrt{n})$$

The answer is $\boxed{0}$ because

$$\lim_{n \to \infty} (\sqrt{n+1} - \sqrt{n}) = \lim_{n \to \infty} \frac{1}{\sqrt{n+1} + \sqrt{n}} = 0$$

Point to note: when n approaches to infinity, a limit can only exist if n is in the denominator. This idea is already used in *Example 2.2.2* on *page 7* where we divide each term in the numerator by n in order to bring n to the denominator. The transformation employed in this example uses a similar idea.

Chapter A: Solutions

Practice 5

Compute
$$\lim_{x \to 4} \frac{3 - \sqrt{x+5}}{x - 4}$$

Multiplying both the denominator and the numerator by $3 + \sqrt{x+5}$ yields

$$\lim_{x \to 4} \frac{3 - \sqrt{x+5}}{x-4} = \lim_{x \to 4} \frac{(3-\sqrt{x+5})(3+\sqrt{x+5})}{(x-4)(3+\sqrt{x+5})}$$
$$= \lim_{x \to 4} \frac{9 - (x+5)}{(x-4)(3+\sqrt{x+5})}$$
$$= \lim_{x \to 4} \frac{-1}{3 + \sqrt{x+5}}$$
$$= \boxed{-\frac{1}{6}}$$

Point to note: in order for the limit to exist when $x \to 4$, the factor $(x-4)$ in the denominator must be canceled. Another point to note: applying the difference of squares formula is a commonly used technique. It was used in the previous practice too.

Practice 6

Find the value of $\lim_{n \to \infty} \sin^2 \left(\pi \sqrt{n^2 + n} \right)$.

(China)

Note that the sine function is periodical and
$$\sqrt{n^2 + n} - n = \frac{n}{\sqrt{n^2 + n} + n}$$

Therefore,
$$\lim_{n \to \infty} \sin^2 \left(\pi \sqrt{n^2 + n} \right)$$

Chapter A: Solutions

$$= \lim_{n\to\infty} \sin\left(\pi\sqrt{n^2+n} - n\pi\right)$$

$$= \lim_{n\to\infty} \sin^2\left(\frac{n\pi}{\sqrt{n^2+n}+n}\right)$$

$$= \lim_{n\to\infty} \sin^2\left(\frac{\pi}{\sqrt{1+\frac{1}{n}}+1}\right)$$

$$= \boxed{1}$$

Practice 7

Use the trigonometric difference to product identity below

$$\sin\alpha - \sin\beta = 2\sin\frac{\alpha-\beta}{2}\cos\frac{\alpha+\beta}{2}$$

and the conclusion of *Example 2.3.1* on *page 8* to evaluate

$$\lim_{\Delta x\to 0}\frac{\sin(x+\Delta x) - \sin x}{\Delta x}$$

Applying the difference to product formula gives:

$$\lim_{\Delta x\to 0}\frac{\sin(x+\Delta x) - \sin x}{\Delta x}$$

$$= \lim_{\Delta x\to 0}\frac{2\sin\frac{\Delta x}{2}\cos\left(x+\frac{\Delta x}{2}\right)}{\Delta x}$$

$$= \lim_{\Delta x\to 0}\frac{\sin\frac{\Delta x}{2}}{\frac{\Delta x}{2}}\cos\left(x+\frac{\Delta x}{2}\right)$$

$$= 1\cdot\cos x$$

$$= \boxed{\cos x}$$

The second last step uses the conclusion of *Example 2.3.1*.

Chapter A: Solutions

Practice 8

Recall
$$\sum_{k=1}^{n} k^3 = \left(\frac{n(n+1)}{2}\right)^2$$

Find the area below the curve $y = x^3$ between $x = 0$ and 1.

Applying the similar model used in solving *Example 2.8* on *page 19* give the desired answer as

$$\lim_{n\to\infty} \sum_{k=1}^{n} \frac{1}{n}\left(\frac{k}{n}\right)^3 = \lim_{n\to\infty} \frac{1}{n^4} \sum_{k=1}^{n} k^3$$
$$= \lim_{n\to\infty} \frac{1}{n^4}\left(\frac{n(n+1)}{2}\right)^2$$
$$= \lim_{n\to\infty} \frac{1}{4}\left(\frac{n+1}{n}\right)^2$$
$$= \boxed{\frac{1}{4}}$$

Practice 9

Use the sandwich theorem to find the value of
$$\lim_{n\to\infty} \sum_{k=1}^{n} \frac{n+k}{n^2+k}$$

Let
$$x_n = \sum_{k=1}^{n} \frac{n+k}{n^2+k}, \quad y_n = \sum_{k=1}^{n} \frac{n+k}{n^2+n}, \quad z_n = \sum_{k=1}^{n} \frac{n+k}{n^2+1}$$

Then, it is clear that $y_n \leq x_n \leq z_n$, and

$$y_n = \sum_{k=1}^{n} \frac{n+k}{n^2+n} = \frac{n^2 + \frac{n(n+1)}{2}}{n^2+n} = \frac{3n+1}{2n+2} \implies \lim_{n\to\infty} y_n = \frac{3}{2}$$

Chapter A: Solutions

$$z_n = \sum_{k=1}^{n} \frac{n+k}{n^2+1} = \frac{n^2 + \frac{n(n+1)}{2}}{n^2+1} = \frac{3n^2+n}{2n^2+2} \implies \lim_{n\to\infty} z_n = \frac{3}{2}$$

Therefore, by the sandwich theorem, we find

$$\lim_{n\to\infty} \sum_{k=1}^{n} \frac{n+k}{n^2+k} = \lim_{n\to\infty} x_n = \boxed{\frac{3}{2}}$$

Points to note: in order to apply the sandwich theorem, we need to relax each term in order to find the upper bound and also to tighten each term to find the lower bound. Meanwhile, in order to simply the sum $\sum_{k=1}^{n}$, it will be ideal if the denominators of those items become the same and are independent of the variable k. Finally, as it is noted in one previous practice that the term with the highest power dominants the growth of a polynomial. This means that changing the denominator (n^2+k) to (n^2+n) or (n^2+1) will not change the value of the limit.

Practice 10

Apply the bounded monotonic function method to show that $e = \lim_{n\to\infty} \left(1 + \frac{1}{n}\right)^n$ exists.

First, let's show that for all $n \geq 1$, it holds

$$\left(1 + \frac{1}{n}\right)^n < \left(1 + \frac{1}{n+1}\right)^{n+1}$$

This can be done by applying the AM-GM inequality on the following $(n+1)$ positive numbers.

$$x_1 = 1, \quad x_2 = x_3 = \cdots = x_{n+1} = 1 + \frac{1}{n}$$

Because they are not all equal, inequality will strictly hold, i.e.

$$\sqrt[n+1]{x_1 x_2 x_3 \cdots x_{n+1}} < \frac{x_1 + x_2 + x_3 + \cdots + x_{n+1}}{n+1}$$

$$\sqrt[n+1]{\left(1+\frac{1}{n}\right)^n} < \frac{1}{n+1}\left(1+n\left(1+\frac{1}{n}\right)\right)$$

$$\left(1+\frac{1}{n}\right)^{\frac{n}{n+1}} < 1+\frac{1}{n+1}$$

$$\left(1+\frac{1}{n}\right)^n < \left(1+\frac{1}{n+1}\right)^{n+1}$$

Next, we are going to show that

$$\left(1+\frac{1}{n}\right)^n < 3$$

Applying binomial expansion gives

$$\left(1+\frac{1}{n}\right)^n = 1 + \binom{n}{1}\frac{1}{n} + \sum_{k=2}^{n}\binom{n}{k}\frac{1}{n^k}$$

$$= 1 + 1 + \sum_{k=2}^{n} \frac{n(n-1)\cdots(n-k+1)}{k!} \cdot \frac{1}{n^k}$$

$$= 2 + \sum_{k=2}^{n} \frac{1}{k!} \cdot \frac{n(n-1)\cdots(n-k+1)}{n^k}$$

$$\leq 2 + \sum_{k=2}^{n} \frac{1}{k!}$$

$$\leq 2 + \sum_{k=2}^{n} \frac{1}{k(k-1)}$$

$$= 2 + \sum_{k=2}^{n} \left(\frac{1}{k-1} - \frac{1}{k}\right)$$

$$= 2 + \left(\frac{1}{2-1} - \frac{1}{n}\right) < 3$$

Therefore, we conclude that $\left(1+\frac{1}{n}\right)^n$ strictly increases with an upper bound. This means that its limit as n approaches infinity must exist. The limit is denoted as the symbol e whose value is approximately 2.718.

Chapter A: Solutions

Practice 11

Show that
$$\lim_{x \to 0} \frac{e^x - 1}{x} = 1$$

Let $h = e^x - 1$, then $x \to 0 \implies h \to 0$. It follows that

$$\lim_{x \to 0} \frac{e^x - 1}{x} = \lim_{h \to 0} \frac{h}{\ln(1+h)}$$
$$= \lim_{h \to 0} \frac{1}{\frac{1}{h}\ln(1+h)}$$
$$= \lim_{h \to 0} \frac{1}{\ln(1+h)^{\frac{1}{h}}}$$

When $h \to 0$, we have $(1+h)^{\frac{1}{h}} = e$ by the definition of e. Therefore, the above relation leads to

$$\lim_{x \to 0} \frac{e^x - 1}{x} = \lim_{h \to 0} \frac{1}{\ln(1+h)^{\frac{1}{h}}} = \frac{1}{\ln e} = 1$$

This result is often used as a steppingstone to solve other problems. Therefore, it should be memorized.

Chapter A: Solutions

A.3 Chapter 3

Practice 1

Find the derivative of a^x where a is a constant.

Let $y = a^x$, then taking logarithm on both sides produces
$$x \ln a = \ln y$$
Differentiating both sides leads to
$$\begin{aligned} \mathrm{d}(x \ln a) &= \mathrm{d} \ln y \\ \ln a (\mathrm{d}x) &= \frac{1}{y} \mathrm{d}y \\ \frac{\mathrm{d}y}{\mathrm{d}x} &= y \ln a \end{aligned}$$

Setting $y = a^x$ back yields
$$\boxed{(a^x)' = a^x \ln a}$$
Note that when $a = e$, this result becomes $(e^x)' = e^x$.

Practice 2

Find the derivative of $\log_a x$ where a is a constant.

The result can be derived from the known relation $(\ln x)' = \frac{1}{x}$.
$$\frac{\mathrm{d}}{\mathrm{d}x} \log_a x = \frac{\mathrm{d}}{\mathrm{d}x} \frac{\ln x}{\ln a} = \frac{1}{\ln a} \left(\frac{\mathrm{d}}{\mathrm{d}x} \ln x \right) = \boxed{\frac{1}{x \ln a}}$$

Practice 3

Find the derivative of function $\arctan x$.

Chapter A: Solutions

Let $g(x) = \arctan x$ and $f(x) = g^{-1}(x) = \tan x$. Then by the inverse function rule, we have

$$g'(x) = \frac{1}{f'(g(x))} = \frac{1}{1+\tan^2(\arctan x)} = \boxed{\frac{1}{1+x^2}}$$

Practice 4

Compute the derivative of $x \ln x$.

Applying the product rule gives

$$(x \ln x)' = x(\ln x)' + (x)' \ln x = x\frac{1}{x} + \ln x = \boxed{1 + \ln x}$$

Practice 5

Convert $x^2 + y^2 = R^2$, where R is a constant, to an explicit form and then compute $\frac{dy}{dx}$. Compare this solution with that presented in *Example 3.7.1* on *page 34*.

The given equation is equivalent to $y = \pm\sqrt{R^2 - x^2}$.

Let $u = R^2 - x^2$, then $y = \pm u^{\frac{1}{2}}$. then

$$\frac{dy}{dx} = \frac{dy}{du} \cdot \frac{du}{dx} = \pm\left(\frac{1}{2}u^{-\frac{1}{2}}\right)(-2x) = \frac{-x}{\pm\sqrt{R^2-x^2}} = -\frac{x}{y}$$

Practice 6

The equation $x^y = y^x$ describes a curve in the first quadrant of the plane containing the point $P = (4, 2)$. Compute the slope of the line that is tangent to this curve at P.

(Bennett)

Chapter A: Solutions

The given equation is equivalent to

$$y \ln x = x \ln y \implies \frac{\ln x}{x} = \frac{\ln y}{y}$$

Taking derivative on both sides gives

$$\frac{\frac{1}{x} \cdot x - \ln x \cdot 1}{x^2} \, dx = \frac{\frac{1}{y} \cdot y - \ln y \cdot 1}{y^2} \, dy$$

$$\frac{1 - \ln x}{x^2} \, dx = \frac{1 - \ln y}{y^2} \, dy$$

$$\frac{dy}{dx} = \frac{1 - \ln x}{1 - \ln y} \cdot \frac{y^2}{x^2}$$

Setting $(x, y) = (4, 2)$ to the above relation gives

$$\frac{dy}{dx} = \frac{1 - \ln 2}{1 - \ln 4} \times \frac{4^2}{2^2} = \boxed{\frac{4(1 - \ln 2)}{1 - 2\ln 2}} \approx -3.177$$

Practice 7

Consider the parabola $y = ax^2 + 2019x + 2019$. There exists exactly one circle which is centered on the x-axis and is tangent to the parabola at exactly two points. It turns out that one of these tangent points is $(0, 2019)$. Find a.

(SMT)

Because the x-coordinate of the parabola's vertex is $-\frac{2019}{2a}$. Therefore, the center of the circle must be $\left(-\frac{2019}{2a}, 0\right)$.

Meanwhile, the slope of the tangent to the parabola equals $2ax + 2019$. Hence, at the point $(0, 2019)$, the slope is 2019. Accordingly, the slope of the radius from the circle center to the tangent point must be $-\frac{1}{2019}$. This means

$$-\frac{0 - 2019}{-\frac{2019}{2a} - 0} = -\frac{1}{2019} \implies a = \boxed{-\frac{1}{4038}}$$

Chapter A: Solutions

Practice 8

Let $f(x)$ be an odd function which is differentiable over $(-\infty, +\infty)$. Show that $f'(x)$ is even.

(UConn)

Because $f(x)$ is odd, therefore

$$f(-x) = -f(x) \implies f(x) = -f(-x)$$

Taking derivative on both sides yields

$$f'(x) = (-f(-x))' = -(f(-x))' = -(-f'(-x)) = f'(-x)$$

The second last step uses the chain rule.

Practice 9

Let $f_0(x) = (\sqrt{e})^x$, and recursively define $f_{n+1}(x) = f'_n(x)$ for integers $n \geq 0$. Compute $\sum_{k=0}^{\infty} f_k(1)$.

(SMT)

Rewrite $f_0(x) = e^{\frac{x}{2}}$. Then $f_1(x) = \frac{1}{2} e^{\frac{x}{2}}$. By induction, we can find that $f_n(x) = \frac{1}{2^n} e^{\frac{x}{2}}$. Hence

$$\sum_{k=0}^{\infty} f(1) = \sum_{k=0}^{\infty} \frac{1}{2^k} e^{\frac{1}{2}} = \sqrt{e} \sum_{k=0}^{\infty} \frac{1}{2^k} = \boxed{2\sqrt{e}}$$

The last step uses the sum of a geometric sequence:

$$\sum_{k=0}^{\infty} \frac{1}{2^k} = 1 + \frac{1}{2} + \frac{1}{2^2} + \frac{1}{2^3} + \cdots = 2$$

Practice 10

Let curve $\mathbb{C}$ is defined as
$$\begin{cases} x = \cot t \\ y = \dfrac{\cos(2t)}{\sin t} \end{cases}$$
where $t \in (0, \pi)$. Find all inflection points of this curve.

(China)

The derivative of this curve can be computed using the implicit derivative method.

$$\frac{dx}{dt} = -\frac{1}{\sin^2 t}$$

$$\frac{dy}{dt} = \frac{-2\sin(2t)\sin t - \cos(2t)\cos t}{\sin^2 t}$$

$$\therefore \quad \frac{dy}{dx} = 2\sin(2t)\sin t + \cos(2t)\cos t$$
$$= 4\sin^2 t \cos t + (1 - 2\sin^2 t)\cos t$$
$$= (1 + 2\sin^2 t)\cos t$$

It follows that

$$\frac{d^2 y}{dx^2} = 4\sin t \cos^2 t - (1 + 2\sin^2 t)\sin t$$
$$= \sin t(4\cos^2 t - 1 - 2\sin^2 t)$$
$$= \sin t(4\cos^2 t - 1 - 2 + 2\cos^2 t)$$
$$= 3\sin t(2\cos^2 t - 1)$$
$$= -3\sin t \cos(2t)$$

Setting $\frac{d^2 y}{dx^2} = 0$ leads to $t_1 = \frac{\pi}{4}$ and $t_2 = \frac{3\pi}{4}$. Meanwhile,

Chapter A: Solutions

- when $0 < t < \frac{\pi}{4}$, $\frac{d^2y}{dx^2} < 0$.
- when $\frac{\pi}{4} < t < \frac{3\pi}{4}$, $\frac{d^2y}{dx^2} > 0$.
- when $\frac{3\pi}{4} < t < \pi$, $\frac{d^2y}{dx^2} < 0$.

Therefore, both points are inflection points. Their corresponding coordinates are $\boxed{(\pm 1, 0)}$.

Practice 11

Compute $\lim\limits_{x \to 0} x \ln x$ and $\lim\limits_{x \to 0} x^x$.

The first expression can be evaluated using the L'Hôpital rule.

$$\lim_{x \to 0} x \ln x = \lim_{x \to 0} \frac{\ln x}{\frac{1}{x}} = \lim_{x \to 0} \frac{\frac{1}{x}}{-\frac{1}{x^2}} = \boxed{0}$$

It follows that

$$\lim_{x \to 0} x^x = \lim_{x \to 0} e^{x \ln x} = e^0 = \boxed{1}$$

Practice 12

Compute

$$\lim_{x \to 0} \frac{(1 - \cos x)^2}{x^2 - x^2 \cos^2 x}$$

(SMT)

First, the limit can be simplified as

$$\lim_{x \to 0} \frac{(1 - \cos x)^2}{x^2 - x^2 \cos^2 x} = \lim_{x \to 0} \frac{(1 - \cos x)^2}{x^2 \sin^2 x} = \left(\lim_{x \to 0} \frac{1 - \cos x}{x \sin x}\right)^2$$

Then, applying the L'Hôpital rule twice leads to

$$\left(\lim_{x \to 0} \frac{\sin x}{\sin x + x \cos x}\right)^2 = \left(\lim_{x \to 0} \frac{\cos x}{\cos x + \cos x - x \sin x}\right)^2 = \boxed{\frac{1}{4}}$$

Practice 13

For a given $x > 0$, let a_n be the sequence defined by $a_1 = x$ for $n = 1$ and $a_n = x^{a_{n-1}}$ for $n \geq 2$. Find the largest x for which $\lim\limits_{n \to \infty} a_n$ exists.

(SMT)

In order for $\lim\limits_{n \to \infty} a_n$ to have a limit L, it must hold that $L = x^L$ or $x = L^{1/L}$. Therefore, the problem is equivalent to maximize $f(L) = L^{1/L}$.

$$\frac{df}{dL} = \frac{d\left(L^{1/L}\right)}{dL} = \frac{d\left(e^{\frac{\ln L}{L}}\right)}{dL} = L^{\frac{1}{L}}\left(\frac{1}{L^2} - \frac{\ln L}{L^2}\right)$$

Setting this to be 0 leads to $L = e$ and accordingly $x = \boxed{e^{1/e}}$

We also need to make sure the value is indeed a maximum. For this, let's check its second derivative at $L = e$.

$$\left.\frac{d^2 f}{dL^2}\right|_e = L^{\frac{1}{L}-4}\left(-3L + \ln^2 L + 2(L-1)\ln L + 1\right)|_e = -e^{1/e-3} < 0$$

Therefore, we find this value is indeed a maximum.

Practice 14

Compute the value of

$$\lim_{x \to \pi} \frac{\ln(2 + \cos x)}{(3^{\sin x} - 1)^2}$$

(China)

Because

$$\lim_{x \to 0} \frac{\ln(1+x)}{x} = 1 \quad \text{and} \quad \lim_{x \to 0} \frac{e^x - 1}{x} = 1$$

Chapter A: Solutions

(The first result can be shown using the L'Hôpital rule. The second result is obtained in last chapter's practice.)

therefore

$$\lim_{x \to \pi} \frac{\ln(2 + \cos x)}{(3^{\sin x} - 1)^2}$$

$$= \lim_{x \to \pi} \frac{\ln(1 + (1 + \cos x))}{(e^{\sin x \ln 3} - 1)^2}$$

$$= \lim_{x \to \pi} \frac{\ln(1 + (1 + \cos x))}{1 + \cos x} \cdot \frac{1 + \cos x}{\sin^2 x \ln^2 3} \left(\frac{\sin x \ln 3}{e^{\sin x \ln 3} - 1} \right)^2$$

$$= \lim_{x \to \pi} \frac{1 + \cos x}{\sin^2 x \ln^2 3}$$

$$= \frac{1}{\ln^2 3} \lim_{x \to \pi} \frac{-\sin x}{2 \sin x \cos x} \quad \text{(L'Hôpital)}$$

$$= \boxed{\frac{1}{2 \ln^2 3}}$$

Practice 15

Let $f(x) = x^2 \cos(ax)$ where a is a constant. Find the 50th order derivative of $f(x)$, i.e. $f^{(50)}(x)$.

(China)

By *(3.25)* on *page 47*, we find

$$f^{(50)} = \sum_{k=0}^{50} (x^2)^{(k)} \cos^{(50-k)}(ax)$$

Because $(x^2)' = 2x$, $(x^2)'' = 2$ and $(x^2)''' = 0$, the above equation has at most 3 terms.

We also note that the derivatives of $\cos x$ alternate among $\pm \sin x$ and $\pm \cos x$:

- $(\cos x)^{(1)} = -\sin x$

- $(\cos x)^{(2)} = -\cos x$
- $(\cos x)^{(3)} = \sin x$
- $(\cos x)^{(4)} = \cos x$

It follows that

- $(\cos x)^{(50)} = -\cos x \implies (\cos(ax))^{(50)} = -a^{50}\cos ax$
- $(\cos x)^{(49)} = -\sin x \implies (\cos(ax))^{(49)} = -a^{49}\sin ax$
- $(\cos x)^{(48)} = \cos x \implies (\cos(ax))^{(48)} = a^{48}\cos ax$

Therefore, the final answer is

$$x^2(-a^{50}\cos(ax)) + \binom{50}{1}(2x)(-a^{49}\sin(ax)) + \binom{50}{2}(2)(a^{48}\cos(ax))$$
$$= -a^{50}x^2\cos(ax) - 100a^{49}x\sin(a) + 2450a^{48}\cos(ax)$$
$$= \boxed{a^{48}\left((2450 - a^2 x^2)\cos(ax) - 100ax\sin(ax)\right)}$$

Practice 16

If water is poured into a right cone whose height is H and base's radius is R at a speed of A, what is the speed the water is rising when the depth of water is half of the cone's height?

Let y be the depth of the water in this cone. Then y is a function of time t. The target of this problem is to compute $\frac{dy}{dt}$ when $y = \frac{H}{2}$.

It is clear that $\frac{dy}{dt}$ is related to the change in the volume of accumulated water V. Meanwhile, V can be calculated as the difference of two cones:

$$V = \frac{1}{3}\pi R^2 H - \frac{1}{3}\pi \left(\frac{R(H-y)}{H}\right)^2 (H-y) = \frac{\pi R^2}{3H^2}\left(H^3 - (H-y)^3\right)$$

Chapter A: Solutions

the change of volume at the moment t can be computed as (note that y is a function of t, thus we must apply the chain rule)

$$\frac{dV}{dt} = \frac{\pi R^2}{H^2}(H-y)^2 \frac{dy}{dt}$$

Apparently, the change in the volume of accumulated water must equal to the speed of which water is poured in, i.e. $\frac{dV}{dt} = A$. Setting this to the relation above yields

$$\frac{dy}{dt} = \frac{AH^2}{\pi R^2(H-y)^2}$$

Finally, we find when $y = \frac{H}{2}$,

$$\boxed{\frac{dy}{dt} = \frac{4A}{\pi R^2}}$$

Practice 17

Use the derivative definition to prove the product rule.

Let $y(x) = f(x)g(x)$. Then by the definition of derivative, we have

$$\begin{aligned}
\frac{d}{dx}y(x) &= \lim_{\Delta x \to 0} \frac{y(x+\Delta x) - y(x)}{\Delta x} \\
&= \lim_{\Delta x \to 0} \frac{f(x+\Delta x)g(x+\Delta x) - f(x)g(x)}{\Delta x} \\
&= \lim_{\Delta x \to 0} \frac{f(x+\Delta x)g(x+\Delta x) - f(x+\Delta x)g(x)}{\Delta x} \\
&\quad + \lim_{\Delta x \to 0} \frac{f(x+\Delta x)g(x) - f(x)g(x)}{\Delta x} \\
&= \lim_{\Delta x \to 0} f(x+\Delta x) \left(\frac{g(x+\Delta x) - g(x)}{\Delta x} \right) \\
&\quad + g(x) \lim_{\Delta x \to 0} \frac{f(x+\Delta x) - f(x)}{\Delta x}
\end{aligned}$$

Now, note that by derivative's definition, we have

$$\lim_{\Delta x \to 0} \frac{g(x + \Delta x) - g(x)}{\Delta x} = g'(x)$$

and

$$\lim_{\Delta x \to 0} \frac{f(x + \Delta x) - f(x)}{\Delta x} = f'(x)$$

Therefore, the original expression equals

$$f(x)g'(x) + g(x)f'(x)$$

Chapter A: Solutions

A.4 Chapter 4

Practice 1

Compute
$$\int \frac{1}{ax+b}\,dx$$

This problem can be solved using the substitution method

$$\int \frac{1}{ax+b}\,dx = \frac{1}{a}\int \frac{1}{ax+b}\,d(ax+b) = \boxed{\frac{1}{a}\ln|ax+b| + C}$$

Practice 2

Let
$$f(x) = \begin{cases} \cos x &, x \in [-\frac{\pi}{2}, 0] \\ e^x &, x \in [0, 1] \end{cases}$$

Compute
$$\int_{-\frac{\pi}{2}}^{1} f(x)\,dx$$

This integral can be done by dividing the interval into two:

$$\int_{-\frac{\pi}{2}}^{1} f(x)\,dx$$
$$= \int_{-\frac{\pi}{2}}^{0} f(x)\,dx + \int_{0}^{1} f(x)\,dx$$
$$= \int_{-\frac{\pi}{2}}^{0} \cos x\,dx + \int_{0}^{1} e^x\,dx$$
$$= \sin x \big|_{-\frac{\pi}{2}}^{0} + e^x \big|_{0}^{1}$$
$$= (0 - (-1)) + (e - 1)$$
$$= \boxed{e}$$

Practice 3

Compute
$$\int_0^4 \frac{dx}{\sqrt{|x-2|}}$$

(SMT)

Split the to-be-integrated interval into two:

$$\int_0^2 \frac{dx}{\sqrt{|x-2|}} = \int_0^2 \frac{dx}{\sqrt{2-x}} = -2\sqrt{2-x}\Big|_0^2 = 2\sqrt{2}$$

$$\int_2^4 \frac{dx}{\sqrt{|x-2|}} = \int_0^2 \frac{dx}{\sqrt{x-2}} = 2\sqrt{x-2}\Big|_2^4 = 2\sqrt{2}$$

The first equation uses substitution $u = 2 - x$ and the second one uses $u = x - 2$. Adding the two results gives the final answer as

$$2\sqrt{2} + 2\sqrt{2} = \boxed{4\sqrt{2}}$$

Practice 4

Evaluate the following integral by using its geometric meaning first. And then compute it again algebraically using the substitution method.
$$\int_0^1 \sqrt{1-x^2}\,dx$$

This integral represent the area of a unit circle in the first quadrant. Therefore, its value should equal $\boxed{\dfrac{\pi}{4}}$.

Meanwhile, this integral can also be evaluated using trigonometric substitution. Let $x = \sin\theta$, then $\theta \in [0, \pi/2]$.

$$\int_0^1 \sqrt{1-x^2}\,dx = \int_0^{\frac{\pi}{2}} \cos\theta\,d\sin\theta$$

Chapter A: Solutions

$$= \int_0^{\frac{\pi}{2}} \cos^2 \theta \, d\theta$$

$$= \int_0^{\frac{\pi}{2}} \frac{1 + \cos 2\theta}{2} \, d\theta$$

$$= \frac{1}{2} \int_0^{\frac{\pi}{2}} (1 + \cos 2\theta) \, d\theta$$

$$= \frac{1}{2} \left(\theta + \frac{1}{2} \sin 2\theta \right) \Big|_0^{\frac{\pi}{2}}$$

$$= \boxed{\frac{\pi}{4}}$$

Practice 5

Compute
$$\int_0^{\frac{\pi}{4}} \frac{1}{\sin x + \cos x} \, dx$$

Applying the trigonometry transformation of
$$a \sin \alpha + b \cos \alpha = \sqrt{a^2 + b^2} \sin(\alpha + \varphi)$$

where φ satisfying $\tan \varphi = \frac{b}{a}$ (see the book *Trigonometry*) gives

$$\int_0^{\frac{\pi}{4}} \frac{1}{\sin x + \cos x} \, dx = \int_0^{\frac{\pi}{4}} \frac{1}{\sqrt{2} \sin\left(x + \frac{\pi}{4}\right)} \, dx$$

Letting $u = x + \frac{\pi}{4}$ and applying the result of *Example 4.7.7* on *page 88* yield:

$$\frac{1}{\sqrt{2}} \int_{-\frac{\pi}{4}}^{0} \frac{1}{\sin u} \, du = \frac{1}{\sqrt{2}} \left(-\ln |\csc u + \cot u| \right) \Big|_{-\frac{\pi}{4}}^{0} = \boxed{\frac{\sqrt{2}}{2} \ln\left(\sqrt{2} + 1\right)}$$

Practice 6

Compute
$$\int \ln x \, dx$$

Chapter A: Solutions

This problem can be solved using substitution and integration by parts. Let $y = \ln x$, then $x = e^y$. It follows that:

$$\int \ln x \, dx = \int y \, de^y = ye^y - \int e^y \, dy = ye^y - e^y + C = \boxed{x \ln x - x + C}$$

This result may be used as a well-known formula when solving more complex problems.

Practice 7

Compute
$$\int x^3 \ln x \, dx$$

This problem can be solved by using the integration by parts method:

$$\int x^3 \ln x \, dx = \frac{1}{4} \int \ln x \, dx^4$$
$$= \frac{1}{4} \left(x^4 \ln x - \int x^4 \, d(\ln x) \right)$$
$$= \frac{1}{4} \left(x^4 \ln x - \int x^3 \, dx \right)$$
$$= \boxed{\frac{1}{4} x^4 \ln x - \frac{1}{16} x^4 + C}$$

Practice 8

Compute
$$\int \frac{1}{\sqrt{a^2 - x^2}} \, dx$$

Similar to *Example 4.4.2* on *page 75*, this problem is better to be solved using regular substitution instead of trigonometric substitution.

$$\int \frac{1}{\sqrt{a^2 - x^2}} \, dx = \int \frac{1}{\sqrt{1 - \left(\frac{x}{a}\right)^2}} \, d\left(\frac{x}{a}\right) = \boxed{\arcsin \frac{x}{a} + C}$$

Chapter A: Solutions

Practice 9

Evaluate
$$\int_0^1 x \arcsin x \, dx$$

(China)

This problem can be solved using the integration by parts method.

$$\int_0^1 x \arcsin x \, dx$$
$$= \int_0^1 \arcsin x \, d\left(\frac{x^2}{2}\right)$$
$$= \frac{x^2}{2} \arcsin x \Big|_0^1 - \int_0^1 \frac{x^2}{2} d(\arcsin x)$$
$$= \frac{\pi}{4} - \frac{1}{2} \int_0^1 \frac{x^2}{\sqrt{1-x^2}} dx$$
$$= \frac{\pi}{4} + \frac{1}{2} \left(\int_0^1 \frac{(1-x^2)-1}{\sqrt{1-x^2}} dx \right)$$
$$= \frac{\pi}{4} + \frac{1}{2} \left(\int_0^1 \sqrt{1-x^2} \, dx - \int_0^1 \frac{1}{\sqrt{1-x^2}} dx \right)$$
$$= \frac{\pi}{4} + \frac{1}{2} \left(\frac{\pi}{4} - \arcsin x \Big|_0^1 \right)$$
$$= \boxed{\frac{\pi}{8}}$$

The third last step uses the result from earlier practice:
$$\int_0^1 \sqrt{1-x^2} \, dx = \frac{\pi}{4}$$

Practice 10

Show that
$$\int_0^\infty \frac{x^2}{1+x^4} dx = \int_0^\infty \frac{1}{1+x^4} dx$$

Let $t = \frac{1}{x}$. Then $dx = -\frac{1}{t^2} dt$. It follows that

$$\int_0^\infty \frac{x^2}{1+x^4} dx = \int_\infty^0 \frac{\frac{1}{t^2}}{1+\frac{1}{t^4}} \left(-\frac{1}{t^2}\right) dt = \int_0^\infty \frac{1}{1+t^4} dt$$

Replacing t with x leads to the conclusion immediately.

Practice 11

Evaluate
$$\int_0^\infty \frac{x^2}{1+x^4} dx$$

By the conclusion of the previous practice, we have

$$\int_0^\infty \frac{x^2}{1+x^4} dx = \frac{1}{2}\left(\int_0^\infty \frac{x^2}{1+x^4} dx + \int_0^\infty \frac{1}{1+x^4} dx\right)$$

Therefore,

$$\int_0^\infty \frac{x^2}{1+x^4} dx = \frac{1}{2}\int_0^\infty \frac{1+x^2}{1+x^4} dx$$
$$= \frac{1}{2}\int_0^\infty \frac{1+\frac{1}{x^2}}{x^2+\frac{1}{x^2}} dx \quad (dividing\ by\ x^2)$$
$$= \frac{1}{2}\int_0^\infty \frac{1}{x^2+\frac{1}{x^2}} d\left(x-\frac{1}{x}\right)$$

Let $u = x - \frac{1}{x}$, then $u \in (-\infty, \infty)$ and the above integral equals

$$\frac{1}{2}\int_{-\infty}^\infty \frac{1}{u^2+2} du = \frac{\sqrt{2}}{4}\int_{-\infty}^\infty \frac{1}{\left(\frac{u}{\sqrt{2}}\right)^2+1} d\left(\frac{u}{\sqrt{2}}\right)$$
$$= \frac{\sqrt{2}}{4} \arctan \frac{u}{\sqrt{2}} \bigg|_{-\infty}^\infty$$
$$= \boxed{\frac{\sqrt{2}}{4}\pi}$$

Chapter A: Solutions

Practice 12

Derive the volume formula of a cone.

Let the radius of the base be r and the height be h. Then the cone can be constructed by rotating the following line about the x-axis.

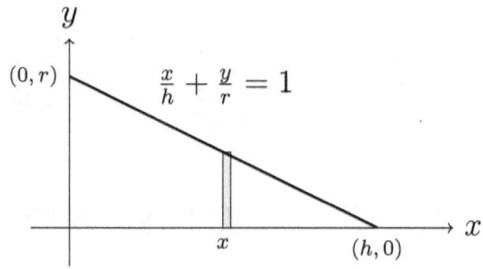

This line's equation can be rewritten as

$$y = -\frac{r}{h}x + r$$

Therefore, its volume can be computed as

$$V = \int_0^h \pi \left(-\frac{r}{h}x + r\right)^2 dx$$

$$= \pi \left(-\frac{h}{r}\right) \int_0^h \left(-\frac{r}{h}x + r\right)^2 d\left(-\frac{r}{h}x + r\right)$$

$$= \pi \left(-\frac{h}{r}\right) \frac{1}{3} \left(-\frac{r}{h}x + r\right)^3 \Big|_0^h$$

$$= \boxed{\frac{1}{3}\pi r^2 h}$$

Practice 13

Construct one polynomial $f(x)$ with real coefficients and with all of the following properties:

a) it is an even function.

b) $f(2) = f(-2) = 0$.

c) $f(x) > 0$ when $-2 < x < 2$.

d) $f(x)$ archives maximum at $x = \pm 1$.

(UConn)

Given $f(x)$ is an even function and constraint (c) above, we know $f(x)$ reaches minimum at $x = 0$. Therefore $f'(x)$ must have at least three zeros at $x = 0$ and $x = \pm 1$. Set $f'(x) = x(x+1)(x-1)$ yields

$$f'(x) = x - x^3 \implies f(x) = \frac{x^2}{2} - \frac{x^4}{4} + C$$

where C is a constant. Solving $f(2) = 0$ gives $C = 2$.

$$\therefore f(x) = \boxed{\frac{x^2}{2} - \frac{x^4}{4} + 2}$$

It can be verified that this polynomial meets all the requirements.

Practice 14

A right circular cone $\mathbb{C}$ has altitude 40 and a circular base of radius 30 inches. A sphere $\mathbb{S}$ is inscribed in $\mathbb{C}$. Find the volume of the region inside $\mathbb{C}$ which is above $\mathbb{S}$.

The answer is $\boxed{300\pi}$.

First, it is easy to find the length of the slant is 50. Next, we claim

Chapter A: Solutions

that the radius of the sphere is 15. This can be computed by taking a vertical cross-section that passes the apex. The triangle's area is 1200 and its semi-perimeter is 80. Hence, the inradius will be $1200 \div 80 = 15$. Computing inradius using area and semi-perimeter is discussed in the book *Geometric Technique*.

Then, using similar triangles, it can be determined that the vertical distance from the apex to the tangent point on the side is 16.

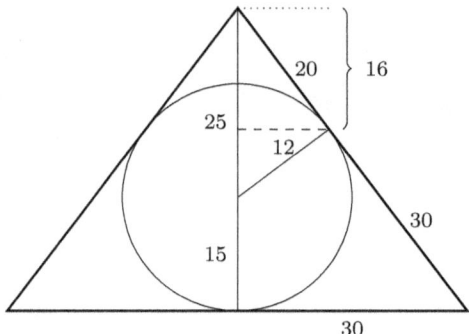

Now, let's put this cross-section in the horizontal direction and set its apex as the origin and its line of symmetry as the x-axis. Then the to-be-determined volume equals the difference between a cone which is constructed by rotating the highlighted triangle and the dome which is constructed by rotating the arc.

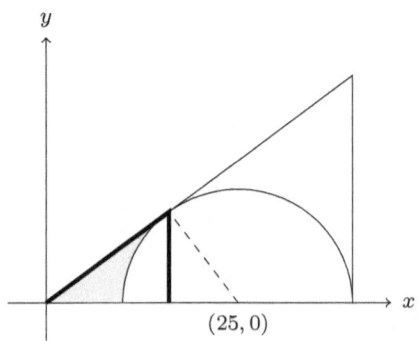

Chapter A: Solutions

The volume of the cone equals

$$\frac{1}{3} \cdot \pi \cdot 12^2 \cdot 16 = 768\pi$$

The equation of the circle is

$$(x-25)^2 + y^2 = 15^2 \implies y^2 = 15^2 - (x-25)^2$$

Therefore the volume of the dome is given by

$$\pi \int_{10}^{16} \left(15^2 - (x-25)^2\right) dx = \pi \left(15^2 x - \frac{1}{3}(x-25)^3\right)\bigg|_{10}^{16} = 468\pi$$

It follows that the final result is $768\pi - 468\pi = \boxed{300\pi}$.

Practice 15

Compute

$$\int_0^\pi \frac{2x \sin x}{3 + \cos^2 x} \, dx$$

(SMT)

The answer is $\boxed{\dfrac{\pi^2}{4}}$. This problem is the same as *Example 4.7.2* on *page 82* after the following transformation:

$$\frac{2x \sin x}{3 + \cos^2 x} = \frac{2x}{3 + (2\cos^2 x - 1)} = \frac{x \sin x}{1 + \cos^2 x}$$

Practice 16

Compute

$$\lim_{x \to 0} \frac{\int_0^x \sin(xt)^2 \, dt}{x^5}$$

(China)

Chapter A: Solutions

Because both the numerator and denominator approach 0 as x approaches 0, applying the L'Hôpital rule gives

$$\lim_{x\to 0} \frac{\int_0^x \sin(xt)^2 dt}{x^5} = \lim_{x\to 0} \frac{\sin x^4}{5x^4} = \boxed{\frac{1}{5}}$$

The transformation of the numerate, i.e.

$$\frac{d}{dx} \int_0^x \sin(xt)^2 \, dt = \sin x^4$$

is based on the fundamental theorem of calculus, *(4.6) on page 72.*

Practice 17

Calculate

$$\lim_{n\to\infty} \frac{1}{n^2} \sum_{k=1}^{n} \left(k \sin \frac{k\pi}{n}\right)$$

(China)

Because

$$\lim_{n\to\infty} \frac{1}{n^2} \sum_{k=1}^{n} \left(k \sin \frac{k\pi}{n}\right) = \lim_{n\to\infty} \frac{1}{n} \sum_{k=1}^{n} \left(\left(\frac{k}{n} \sin \frac{k}{n}\pi\right)\right)$$

therefore the desired result is the Riemann integral of function $f(x) = x \sin(x\pi)$ over $[0, 1]$, i.e.

$$\int_0^1 x \sin(x\pi) dx = \frac{1}{\pi} \int_0^1 x \sin(x\pi) d(x\pi)$$
$$= -\frac{1}{\pi} \int_0^1 xd\cos(x\pi)$$
$$= -\frac{1}{\pi} \left(x \cos(x\pi)\big|_0^1 - \int_0^1 \cos(x\pi) dx\right)$$
$$= -\frac{1}{\pi} \left(-1 - \frac{1}{\pi} \sin(x\pi)\Big|_0^1\right)$$
$$= \boxed{\frac{1}{\pi}}$$

Chapter A: Solutions

Practice 18

Evaluate $\lim_{n \to \infty} S_n$ where

$$S_n = 1 - \frac{1}{2} + \frac{1}{3} - \frac{1}{4} + \frac{1}{5} - \cdots + (-1)^{n-1}\frac{1}{n}$$

We first show that
$$\lim_{n \to \infty} S_{2n} = \ln 2$$

Then because
$$\lim_{n \to \infty} S_{2n+1} = \lim_{n \to \infty} \left(S_{2n} + \frac{1}{2n+1} \right) = \lim_{n \to \infty} S_{2n} = \ln 2$$

therefore we can conclude that
$$\lim_{n \to \infty} S_n = \boxed{\ln 2}$$

To show $\lim_{n \to \infty} S_{2n} = \ln 2$, let's first transform S_{2n}:

$$\begin{aligned}
S_{2n} &= 1 - \frac{1}{2} + \frac{1}{3} - \frac{1}{4} + \cdots - \frac{1}{2n} \\
&= 1 + \frac{1}{2} + \frac{1}{3} + \cdots + \frac{1}{2n} - 2\left(\frac{1}{2} + \frac{1}{4} + \cdots + \frac{1}{2n}\right) \\
&= 1 + \frac{1}{2} + \frac{1}{3} + \cdots + \frac{1}{2n} - \left(\frac{1}{1} + \frac{1}{2} + \cdots + \frac{1}{n}\right) \\
&= \frac{1}{n+1} + \frac{1}{n+2} + \cdots + \frac{1}{2n} \\
&= \frac{1}{n}\left(\frac{1}{1+\frac{1}{n}} + \frac{1}{1+\frac{2}{n}} + \cdots + \frac{1}{1+\frac{n}{n}}\right)
\end{aligned}$$

Applying the rectangular approximation model, the last expression is an approximation of $\frac{1}{1+x}$ between $(0, 1)$. Therefore,

$$\lim_{n \to \infty} S_{2n} = \int_0^1 \frac{1}{1+x}\, dx = \int_0^1 \frac{1}{1+x}\, d(x+1) = \ln(x+1)\big|_0^1 = \ln 2$$

Practice 19

Compute
$$I = \iiint_S \frac{dxdydz}{(1+x+y+z)^2}$$
where $S = \{x \geq 0, y \geq 0, z \geq 0, x+y+z \leq 1\}$.

(UConn)

First, note that $0 \leq z \leq 1-(x+y)$ where $(x,y) \in D = \{x \geq 0, y \geq 0, x+y \leq 1\}$. Then

$$I = \iint_D \left(\int_0^{1-(x+y)} \frac{1}{(1+x+y+z)^2} dz \right) dxdy$$

$$= \iint_D \left(-\frac{1}{1+x+y+z} \Big|_0^{1-(x+y)} \right) dxdy$$

$$= \iint_D \left(\frac{1}{1+x+y} - \frac{1}{2} \right) dxdy$$

Now, noting that $y \in [0, 1-x]$ and $x \in [0,1]$ leads to

$$I = \int_0^1 \left(\int_0^{1-x} \left(\frac{1}{1+x+y} - \frac{1}{2} \right) dy \right) dx$$

$$= \int_0^1 \left(\left(\ln 1+x+y - \frac{1}{2}y \right) \Big|_0^{1-x} \right) dx$$

$$= \int_0^1 \left(\ln 2 - \ln(1+x) - \frac{1}{2}(1-x) \right) dx$$

$$= \left((\ln 2)x - ((1+x)\ln x - x) - (\frac{1}{2}x - \frac{1}{4}x^2) \right) \Big|_0^1$$

$$= \boxed{-\ln 2 + \frac{3}{4}}$$

Chapter A: Solutions

Practice 20

Let $f : \mathbb{R} \to \mathbb{R}$ be a periodic continuous function of period $T > 0$, that is $f(x+T) = f(x)$ holds for any $x \in \mathbb{R}$. Show that

$$\lim_{x \to \infty} \frac{1}{x} \int_0^x f(t)dt = \frac{1}{T} \int_0^T f(t)dt$$

For any $x > 0$, there exist a unique pair of $k > 0$ and $0 \leq a < T$ such that $x = kT + a$. As $x \to \infty$, so will $k \to \infty$. It follwos

$$\frac{1}{x} \int_0^x f(t)dt = \frac{1}{kT+a} \int_0^{kT+a} f(t)dt$$
$$= \frac{1}{kT+a} \left(\int_0^{kT} f(t)dt + \int_{kT}^{kT+a} f(t)dt \right)$$

Because $f(t)$ is periodic, thus

$$\int_0^{kT} f(t)dt = k \int_0^T f(t)dt \quad \text{and} \quad \int_{kT}^{kT+a} f(t)dt = \int_0^a f(t)dt$$

Hence,

$$\frac{1}{x} \int_0^x f(t)dt = \frac{k}{kT+a} \int_0^T f(t)dt + \frac{1}{kT+a} \int_0^a f(t)dt$$

Making $k \to \infty$ yields the desired result.

Practice 21

Without explicitly evaluating the integral, show that

$$\lim_{n \to \infty} \int_1^2 \ln^n x \, dx = 0 \quad \text{and} \quad \lim_{n \to \infty} \int_2^3 \ln^n x \, dx = \infty$$

We note that function $\ln x$ is monotonically increasing. Because the interval is $[1, 2]$, hence $\ln x \leq \ln 2 < 1$. It follows that

$$0 \leq \lim_{n \to \infty} \int_1^2 \ln^n x \, dx \leq \lim_{n \to \infty} \int_1^2 \ln^n 2 \, dx = 0$$

Chapter A: Solutions

In order to prove the second relation, it is sufficient to show that part of the integral is already divergent. Let $2 < a = 1.0001e < 3$. Then

$$\lim_{n\to\infty} \int_2^3 \ln^n x \, dx > \lim_{n\to\infty} \int_a^3 \ln^n x \, dx > \lim_{n\to\infty} \int_a^3 \ln^n a \, dx$$

Now because $\ln a > 1$, therefore $\ln^n a$ diverges. It follows that the last expression diverges.

Practice 22

Determine the differentiable function $f(x)$ such that

$$f(x)\cos x + 2\int_0^x f(t)\sin t \, dt = x + 1$$

(China)

Differentiating both sides of the given equation and noting *Equation 4.6* on *page 72* yield

$$f'(x)\cos x - f(x)\sin x + 2f(x)\sin x = 1$$
$$\therefore \quad f'(x) + f(x)\tan x = \frac{1}{\cos x}$$

This is a standard equation which can be solved using the integrating factor method. Here, the integrating factor equals

$$u(x) = e^{\int \tan x \, dx} = e^{\int \frac{\sin x}{\cos x} dx} = e^{-\int \frac{1}{\cos x} d\cos x} = e^{-\ln \cos x} = \frac{1}{\cos x}$$

Setting this to the solution formula *(4.29)* on *page 97* gives

$$f(x) = \cos x \left(\int \frac{1}{\cos^2 x} dx + C \right) = \cos x \left(\tan x + C \right) = \sin x + C\cos x$$

where C is a constant. Now, setting $x = 0$ to the originally given equation leads to $f(0) = 1$. Therefore C must equal 1. Hence, the final answer is

$$f(x) = \boxed{\sin x + \cos x}$$

A.5 Chapter 5

Practice 1

Expand $f(x) = x^4 - 2x^3 + 1$ around $x_0 = 2$.

As the degree of $f(x)$ is 4, its fifth order derivative and above will be zero.

$$f(x) = x^4 - 2x^3 + 1 \implies f(2) = 1$$
$$f'(x) = 4x^3 - 6x^2 \implies f'(2) = 8$$
$$f''(x) = 12x^2 - 12x \implies f''(2) = 24$$
$$f'''(x) = 24x - 12 \implies f'''(2) = 36$$
$$f''''(x) = 24 \implies f''''(2) = 24$$

Therefore, the desired expansion equals

$$f(x) = 1 + 8(x-2) + \frac{24}{2!}(x-2)^2 + \frac{36}{3!}(x-2)^3 + \frac{24}{4!}(x-2)^4$$
$$= \boxed{1 + 8(x-2) + 12(x-2)^2 + 6(x-2)^3 + (x-2)^4}$$

Practice 2

Estimate the value of $\sqrt[4]{10018}$.

This problem can be solved by employing *(5.17)* on *page 137*.

$$\sqrt[4]{10018} = (10000 + 18)^{\frac{1}{4}} = 10 \times \left(1 + \frac{18}{10000}\right)^{\frac{1}{4}}$$

The last term can be estimated as

$$1 + \frac{1}{4} \times \frac{18}{10000} = 1.00045$$

Therefore we conclude

$$\sqrt[4]{10018} \approx 10 \times 1.00045 = \boxed{10.0045}$$

Chapter A: Solutions

Practice 3

Let $f(x)$ be a twice differentiable continuous function, and $f(0) = f'(0) = 0$, $f''(0) = 6$. Find the value of

$$\lim_{x \to 0} \frac{f(\sin^2 x)}{x^4}$$

(China)

Expand $f(x)$ around $x = 0$ gives

$$f(x) = f(0) + f'(0)x + \frac{1}{2}f''(0)x^2 + O\left(x^3\right) = 3x^2 + O\left(x^3\right)$$

Therefore,

$$f(\sin^2 x) = 3\sin^4 x + O(\sin^6 x)$$

This means

$$\lim_{x \to 0} \frac{f(\sin^2 x)}{x^4} = \lim_{x \to 0} \frac{3\sin^4 x + O(\sin^6 x)}{x^4} = \boxed{3}$$

Practice 4

What is the smallest natural number n for which the following limit exists?

$$\lim_{x \to 0} \frac{\sin^n x}{\cos^2 x (1 - \cos x)^3}$$

(SMT)

First, because

$$\lim_{x \to 0} \frac{\sin^n x}{\cos^2 x (1 - \cos x)^3} = \lim_{x \to 0} \frac{\sin^n x}{(1 - \cos x)^3}$$

therefore, we can safely remove $\cos^2 x$ in the denominator from consideration. Now, by Taylor expansion, we have

$$\sin x = x + O(x^3) \quad \text{and} \quad (1 - \cos x) = \frac{x^2}{2!} + O(x^4)$$

182

It follows that
$$\lim_{x\to 0}\frac{\sin^n x}{(1-\cos x)^3} = \lim_{x\to 0}\frac{x^n}{(0.5x^2)^3}$$

For the last limit to exist, the exponent of x in the numerator must be at lease equal to that in the denominator. Hence, the answer is $\boxed{6}$.

Practice 5

Compute
$$\lim_{x\to 0}\frac{\frac{x^2}{2}+1-\sqrt{1+x^2}}{(\cos x - e^{x^2})\sin(x^2)}$$

(China)

By Taylor's expansion, we have
$$\sqrt{1+x^2} = 1 + \frac{x^2}{2} - \frac{x^4}{8} + R\left(x^4\right)$$

Therefore, the numerator of the given expression equals
$$\frac{x^2}{2}+1-\sqrt{1+x^2} = \frac{1}{8}x^4 + R(x^4)$$

Similarly, the denominator equals
$(\cos x - e^{x^2})\sin(x^2)$
$$= \left(\left(1 - \frac{x^2}{2!} + R\left(x^2\right)\right) - \left(1 + x^2 + R\left(x^2\right)\right)\right)\left(x^2 + R\left(x^2\right)\right)$$
$$= \left(-\frac{3}{2}x^2 + R\left(x^2\right)\right)\left(x^2 + R\left(x^2\right)\right)$$
$$= -\frac{3}{2}x^3 + R\left(x^4\right)$$

It follows that
$$\lim_{x\to 0}\frac{\frac{x^2}{2}+1-\sqrt{1+x^2}}{(\cos x - e^{x^2})\sin(x^2)} = \lim_{x\to 0}\frac{\frac{1}{8}x^4 + R\left(x^4\right)}{-\frac{3}{2}x^4 + R\left(x^4\right)} = \boxed{-\frac{1}{12}}$$

Chapter A: Solutions

Practice 6

Determine if the series $\{\frac{n}{e^n}\}$ converges.

This series converges because

$$\lim_{n\to\infty} \left| \frac{n+1}{e^{n+1}} \div \frac{n}{e^n} \right| = \lim_{n\to\infty} \frac{n+1}{n} \cdot \frac{1}{e} = \frac{1}{e} < 1$$

Practice 7

Show that the inequality $x > \ln(x+1)$ holds for $x > 0$:

The to-be-proved inequality is equivalent to $e^x > x+1$ which can be proved using Taylor expansion. This is because

$$e^x = 1 + x + \frac{x^2}{2!} + \frac{x^3}{3!} + \cdots$$

and all terms are positive.

Practice 8

Is $\sum_{n=1}^{\infty} \frac{1}{\ln(n+1)}$ convergent?

The answer is no. By the conclusion of the previous practice, we have

$$n > \ln(n+1) \implies \frac{1}{\ln(n+1)} > \frac{1}{n}$$

Because $\sum_{n=1}^{\infty} \frac{1}{n}$ diverges, therefore the given series diverges as well.

Chapter A: Solutions

Practice 9

Compute the limit of the power series below as a rational function in x:

$$1\cdot 2 + (2\cdot 3)x + (3\cdot 4)x^2 + (4\cdot 5)x^3 + (5\cdot 6)x^4 + \cdots, \qquad (|x| < 1)$$

(UConn)

Because
$$\frac{1}{1-x} = 1 + x + x^2 + x^3 + \cdots$$
is convergent, therefore it is possible to take derivative on both sides. This gives
$$\frac{1}{(1-x)^2} = 1 + 2x + 3x^2 + 4x^3 + \cdots$$

Taking derivative on both sides again yields:
$$\frac{2}{(1-x)^3} = 2 + (2\cdot 3)x + (3\cdot 4)x^2 + (4\cdot 5)x^3 + \cdots$$

Therefore, the answer is $\boxed{\dfrac{2}{(1-x)^3}}$.

Practice 10

Determine the values of α and β such that
$$\lim_{n\to\infty} \frac{n^\alpha}{n^\beta - (n-1)^\beta} = 2020$$

Dividing both the denominator and the numerator by n^β and then expanding the denominator give

$$\frac{n^\alpha}{n^\beta - (n-1)^\beta} = \frac{n^{\alpha-\beta}}{1 - \left(1 - \frac{1}{n}\right)^\beta} = \frac{n^{\alpha-\beta}}{1 - \left(1 - \frac{\beta}{n} + O\left(\frac{1}{n^2}\right)\right)} = \frac{n^{\alpha-\beta+1}}{\beta - O\left(\frac{1}{n}\right)}$$

Chapter A: Solutions

Because
$$\lim_{n\to\infty} O\left(\frac{1}{n}\right) = 0$$
we have
$$\lim_{n\to\infty} \frac{n^\alpha}{\beta - (n-1)^\beta} = \begin{cases} 0 & , \alpha - \beta + 1 < 0 \\ \frac{1}{\beta} & , \alpha - \beta + 1 = 0 \\ \infty & , \alpha - \beta + 1 > 0 \end{cases}$$

Comparing this result with the given conditions will lead to the conclusion that $2020 = \frac{1}{\beta}$ and $\alpha - \beta + 1 = 0$.

$$\therefore \quad (\alpha, \beta) = \boxed{\left(-\frac{2019}{2020}, \frac{1}{2020}\right)}$$

Practice 11

Prove the absolute value test method, i.e. if $\{|a_n|\}$ converges, so will $\{a_n\}$.

First, we note that
$$0 \leq \sum_{n=1}^{\infty}(a_n + |a_n|) \leq \sum_{n=1}^{\infty} 2|a_n|$$

Therefore, by the comparison test, we find the series $\sum_{n=1}^{\infty}(a_n + |a_n|)$ converges. It follows that the series
$$\sum_{n=1}^{\infty} a_n = \sum_{n=1}^{\infty}(a_n + |a_n|) - \sum_{n=1}^{\infty} |a_n|$$
must converge because it is a difference of two convergent series.

Practice 12

Prove the ratio test, i.e. if $\lim_{n\to\infty}\left|\frac{a_{n+1}}{a_n}\right| = L$ and $L < 1$, then the series $\{a_n\}$ converges absolutely

Let $r = \frac{L+1}{2}$, then $L < r < 1$ because $L < 1$.

Meanwhile, because $\lim_{n\to\infty}\left|\frac{a_{n+1}}{a_n}\right| = L$, therefore there exists a sufficiently large N such that for $n > N$, the relation

$$\left|\frac{a_{n+1}}{a_n}\right| < r \implies |a_{n+1}| < r|a_n|$$

always holds by the definition of limits. This relation implies that $|a_{n+k}| < r^k|a_n|$ for all $n > N$ and $k > 0$. It follows that

$$\sum_{k=N+1}^{\infty}|a_k| = \sum_{k=1}^{\infty}|a_{N+k}| < \sum_{k=1}^{\infty}r^k|a_N| = |a_N|\sum_{k=1}^{\infty}r^k = |a_N|\frac{r}{1-r}$$

because $0 < r < 1$. Hence, we find $\sum_{k=N+1}^{\infty}|a_k|$ is an increasing sequence with an upper bound which means it must converge.

Practice 13

Determine whether or not these two series converge:

$$(A)\ \sum_{n=1}^{\infty}\sin\left(\frac{\cos n}{n^2}\right) \qquad (B)\ \sum_{n=1}^{\infty}\cos\left(\frac{\sin n}{n^2}\right)$$

(Bennett)

The first series converges, but the second one diverges.

Because $\sin x \leq x$ holds for all $x \in \left[0, \frac{\pi}{2}\right]$ and the sine function is odd, therefore we have $|\sin x| \leq |x|$ hold for all $x \in \left[-\frac{\pi}{2}, \frac{\pi}{2}\right]$. Then,

Chapter A: Solutions

as the sine function is periodic, it can be shown that $|\sin x| \leq |x|$ for all x. It follows that

$$0 \leq \sum_{n=1}^{\infty} \left|\sin\left(\frac{\cos n}{n^2}\right)\right| \leq \sum_{n=1}^{\infty} \left|\frac{\cos n}{n^2}\right| \leq \sum_{n=1}^{\infty} \left|\frac{1}{n^2}\right|$$

Now, because $\sum_{n=1}^{\infty} \left|\frac{1}{n^2}\right|$ is convergent, hence series (A) converges absolutely and thus converges.

For the second series diverges because

$$\lim_{n \to \infty} \frac{\sin n}{n^2} = 0 \implies \lim_{n \to \infty} \cos\left(\frac{\sin n}{n^2}\right) = 1 > 0$$

Practice 14

It is well-known that the solution to the Fibonacci sequence is

$$F_n = \frac{1}{\sqrt{5}}\left(\left(\frac{1+\sqrt{5}}{2}\right)^n - \left(\frac{1-\sqrt{5}}{2}\right)^n\right)$$

Show that

$$\lim_{n \to \infty} \frac{F_{n+1}}{F_n} = \frac{1+\sqrt{5}}{2}$$

Let

$$\varphi = \frac{1+\sqrt{5}}{2} \implies -\varphi^{-1} = \frac{1-\sqrt{5}}{2}$$

Hence

$$\lim_{n \to \infty} \frac{F_{n+1}}{F_n} = \frac{\varphi^{n+1} - (-\varphi)^{-(n+1)}}{\varphi^n - (-\varphi)^{-n}} = \lim_{n \to \infty} \frac{\varphi + \frac{(-1)^{n+1}}{\varphi^{2n+1}}}{1 + \frac{(-1)^n}{\varphi^{2n}}}$$

Now both $\frac{(-1)^{n+1}}{\varphi^{2n+1}}$ and $\frac{(-1)^n}{\varphi^{2n}}$ are alternating sequences. By the Leibniz test, both of them approach 0 because $\lim_{n \to \infty} \frac{1}{\varphi^n} = 0$. Setting

Chapter A: Solutions

this result back to the previous equation leads to the desired results immediately.

Practice 15

The oscillation is the time for a pendulum to complete one full back-and-forth movement. Its period T can be computed using the following formula where L is the length of the pendulum and $g \approx 9.8 m/s^2$ is the acceleration of gravity:

$$T = 2\pi \sqrt{\frac{l}{g}}$$

If a particular pendulum's period is 1s, what will be the discrepancy per day if its length is reduced by 1cm due to temperature change?

Differentiating the given formula yields

$$dT = \frac{2\pi}{\sqrt{g}} \frac{1}{2\sqrt{l}} dl \implies \Delta T = \frac{\pi}{\sqrt{g}\sqrt{l}} \Delta l$$

Because the normal period is 1, therefore

$$1 = 2\pi \sqrt{\frac{l}{g}} \implies l = \frac{g}{4\pi^2}$$

Setting this to the previous relation and also $\Delta l = -0.0001$m:

$$\Delta T = \frac{2\pi^2}{g} \Delta l \approx -0.0002s$$

Because there are totally $24 \times 60 \times 60 = 86400$ seconds in a day, therefore this pendulum will become $86400 \times 0.0002 = \boxed{17.28}$ seconds faster every day.

Chapter A: Solutions

www.ingramcontent.com/pod-product-compliance
Lightning Source LLC
Chambersburg PA
CBHW021815170526
45157CB00007B/2595